Lean Modeling for Engineers

DLR Associates Series

DAN RYAN
DLR ASSOCIATES

authorHOUSE®

AuthorHouse™
1663 Liberty Drive
Bloomington, IN 47403
www.authorhouse.com
Phone: 1-800-839-8640

No part of this book may be reproduced, stored in a retrieval system, or transmitted by any means without the written permission of the author.

First published by AuthorHouse 1/18/2010

ISBN: 978-1-4490-7032-8 (sc)
ISBN: 978-1-4490-7033-5 (e)

Printed in the United States of America
Bloomington, Indiana

This book is printed on acid-free paper.

Preface

This book reflects the many changes that advanced computer technology has recently brought to engineering design and manufacturing as practiced by an engineer or technician. The purpose of *Lean Modeling for Engineers* is to provide the reader with a comprehensive, practical guide that will aid both the practicing professional and the student in mastering the essentials of producing design models from a laptop based graphics workstation.. The text material is designed to introduce the reader to each phase of the subject, step-by-step to enable one to use the various pieces of the graphics workstation.

The current state-of-the-art for producing lean models to study design parameters has definite obligations for the professional person. One of the obligations is the use of the graphics workstation in the production of three-dimensional shapes, wireforms, solids, surface specifications and diagrams. This book reflects the many requests that have come from those who have used my books published by Marcel Dekker Inc and CRC Press. These requests have come from the thousands of readers who have used chapters to introduce design modeling. The theory of presentation in this book begins with a CAD Package called AME (Advanced Modeling Extension) and ends with a complete understanding of what lean modeling is and how it can be studied as a separate discipline - that is, the beginning level has not changed - but the applications and illustrations have all been updated to reflect the reader requests for a wide range of hardware platforms (graphic workstations) from many different manufacturers that will run several sets of software options from various software producers. These options are used whenever possible.

Lean Modeling for Engineers is tailored for a continuing engineering education seminar. For distance learners or short, technical seminars, the chapters dealing with LISP computer program extensions my be omitted without destroying the continuity of the book. The selection of material in this book is based on the premise that the reader has not had a course in computer-aided engineering (CAE) or computer-aided manufacturing (CAM) techniques. Therefore, many basic situations are included with explanations or solutions. This makes if possible to keep the emphasis on lean design modeling, not drafting of pictorials. Methods of using existing CAD software is stressed, as is the procedure for writing new LISP programs. An ability to create new software functions can best be developed after the basic lean modeling techniques are well understood, however.

The unique features of this book are:

1. Lean Modeling - using CAD software,
2. Workstation introduction of lean model presentations,
3. Common lean modeling techniques,
4. Industrial applications including lean dimensioning, tolerancing, surface control specification and manufacture.

The author makes no claim to originality of the items used or the functions selected. This book has drawn heavily on his experiences teaching continuing engineering education and writing distance learning packages. Whenever he found a "short cut" for the engineering design processing, he placed it in a three ring notebook. This book is the "best of the notebook". The unique character of this book lies in its industrial orientation, the sequencing of the topics which are discussed and the documentary and user-oriented manner in which they are presented.

DAN RYAN
Professor Emeritus
Clemson University

Contents

LEAN MANUFACTURING MODELS

1

Introduction: Lean Modeling

This chapter is an outline of the remainder of this reference book, giving you a brief overview of the following chapters so that you can learn what a lean model is and how they streamline the engineering design process. You can use this reference to concentrate on the building blocks necessary to implement the process. The technique of using a lean process to build a model is a powerful one. The use of a laptop computer, wireless routing and portability, to speed up the design process is absolutely critical as the United States competes in the twenty-first century.

I would encourage you, the reader, to skip over those chapters that you have already completed in your study of lean productivity analyses. This reference book was completed after fifteen years of consulting and thirty years of teaching engineering at Clemson University. Whenever I found a "short cut" or a lean process for engineers, I put it in a large three ring notebook. This publication is the "best of the notebook".

1.1 Lean Modeling for Engineers

Each field of engineering: Architectural, Aeronautical, Automotive, Civil, Ceramic, Computer, Electrical, Industrial, Mechanical, Nuclear, and Systems engineering all can use lean modeling. The usage ranges from; parts manufacture, assembly processes, remote telemanipulation, space research, welding, palletizing and material handling.

This book will show you a way to produce acceptable lean models in the framework of high productivity. In order to do this, certain standards have been be set. Standards are meant to offer a common ground for both the development of skills and the interfacing of these skills to computer hardware. The choice of hardware for this book is described in detail in Chapter 3, for now it is important to know that any type of laptop can be used with this book. This book is hardware and software independent. The software products described in Chapter 3 are given as possible choices for use on the hardware chosen. I have changed laptop hardware manufacturers and software supplies, five times over the past fifteen years. I started with Gateway, moved to DELL, tried HP and Sony and I am back with HP who sell the Compaq. I have found that the cost of the laptop has very little to do with how it will function in the design of a product. Now I recommend that readers purchase the least expensive laptop with largest memory that they can find. The last group of laptops purchased for a client were from Compaq at a cost of less than 400 dollars each. For those readers very familiar with hardware and software this may seen trite, either bear with us or skip those chapters. While the pace of the book is slow, it is steady and it introduces a new technology to the reader who is willing to understand and apply lean modeling.

1.2 Lean Modeling Documentation

Chapter two's purpose is to introduce lean modeling as a design process used in industry. The following items are covered: 1) typical hardware used, 2) flow charted steps in design process, 3) example robotics problem, 4) brainstorm session, idea charts, and test analysis, 5) refinement, lean model layout, and display, and 6) lean testing, simulation and modeling.

When I teach a seminar, I find that people do not want a lot of pretty pictures or "computer science" thrown at them. They ask, "Where is the on and off switch? Do you have any examples of the 'lean' process to show us? How much of a savings will I have if I use this? Can I design something, right now?" I brought my laptop with me, can I download some of your examples to take back to my boss? Are your class notes available as any e-book?" The answers are always the same, "You already know where the on/off switch is, it is on your laptops. You are connected to the class room wireless router and you can access anything you see listed on your welcome screen. I hope to get these notes into a book form so that you will be able to download it as an e-book or get it in paper back from the same publisher." There are no illustrations in this book, they are all contained in a single directory on a separate CD. The CD is free by contacting: DLR Associates, P.O. Box 11, Sunset, SC 29685

1.3 Lean Model Automation

Professional engineers produce much of their lean modeling at laptops connected to a workstation controller. These types of workstations contain the following: (1) laptop computer, (2) scanner, and (3) graphics tablet. Several of these portable, laptop workstations can be connected to a single large plotter or printer. It would not be cost effective to purchase printers or plotters for individual, portable laptops. Many engineering offices also connect each laptop to a much larger processor which is used as a system controller. All the models, sketches, notes, bills of materials and the like are stored inside the controller. Usually, nothing is stored at the laptop workstation. This is also cost effective because a single copy of the office software is purchased for the controller, not multiple copies for laptops.

For individual, one man offices, like a small architectural office, I would not recommend a network as described above. I would think that you could buy an external hard drive for the office space. Place the software on this device and use a "docking" arrangement with the laptop. When you are out of the office, with a client or on the job site, input your ideas, notes and other items so you can connect later when you are back in the office.

1.4 Notation and Surface Descriptions

Chapter 4 has a unique approach to lettering and other notation for surface descriptions, stylized solely for the production of lean model surfaces. The styles of lettering and surface notation provided by software providers are shown in Chapter 4, one of these may be selected as the default office acceptable lettering that appears on lean models. In addition to the default lettering style, special shape fonts can also be formed. After a special font style is made, however, the software has the ability to generate this symbol. The purpose of this chapter then is to develop a system which will consist of the laptop requirements in both hardware and software instructions. Then we will describe how to use the equipment, and how to code the surface description fonts for storage in the system controller.

1.5 Sketching Techniques for Lean Modeling

Much of the lean model construction in the first four chapters was in **SKETCH** format. Whenever you see a bold, capital letter used in this book, it means that it is stored in the controller and is usually the location for a modeling directory, software function, or display command. You type it from the laptop and the controller branches to that portion of the software. Think of the software as a story book. Each chapter in the book has a title like SKETCH.

A **DLR DIRECTORY** sketch is not a freehand sketch. It is an automated format for using a simple quick method for capturing ideas. It would be impractical to design anything by starting with a DLR software generated lean model before preliminary concepts are discussed and approved. It would also be even more difficult to design from purely verbal descriptions. The research and planning for a lean model include idea sketches for communication, preliminary idea diagrams, notes, dimensions, and specifications. All these will be needed later in the final preparation of a lean model. Freehand model sketching is introduced in this chapter directly after automated notation so that the quality of the lettering, sketching and lean model will be compatible. When you finish this chapter, you should be able to model curved as well as straight line parts with freehand circular holes, rounds, fillets and curved surfaces. Your lean sketches will be similar to those described in Chapter 5. This chapter contains sketching instructions to create a working database. This database may be obtained by the use of a graphics tablet attached to the USB port of the laptop or by wireless router. In this chapter we use the graphics tablet and pointing device just like a lead pencil and engineering cross-section paper. Throughout the material in this chapter the concept of a laptop together with CAD software is stressed, beginning with the graphics tablet and CAD software freehand sketching commands and moves on to the other software packages. The chapter ends with the use of the laptop for the building of a sketching library stored as **BLOCKS.**

1.6 Lean Model Image Processing

Lean image processing applications is generally thought of as being either passive (use a scanner) or active (use online images) in nature. When an engineer selects an online mode rather than manual documentation that is scanned, a decision is made if the lean models can be displayed from software in real-time. An image that is scanned is referred to as computer-assisted rather than computer-aided. The two terms aided and assisted do not carry the same meaning. Computer-aided is an active process because it appears that the laptop has the ability to solve image problems without prior input by humans. Types of hardware are referred to as dumb or smart. A scanner is dumb, it copies whatever it sees manually drawn or sketched and places it in a display **file.dxf**. A smart device stores display files as **.dwg, .skd, .sld,.slb, or .scr.**

Computer-assisted, passive, dumb hardware devices may be used as much as computer-aided, active, smart devices in a lean environment. If model project files are being developed during the early, preliminary stages of a design, dumb devices are used more. In the final stages of project documentation smart devices are used more. In the last few years real-time graphics and design software packages for engineers have been used. This book makes use of a DLR software package called **AME** which is the application software for mechanical, industrial and electrical engineers. The AME package is written especially for active graphics applications falling into one of four major types:1) wireform, frame or mesh; see Chapter 10, 2) 3-D surface; see Chapter 11, 3) solids; see chapter 12, or 4) section and profile, see Chapter 13.

1.7 Storing Lean Models

The basic technique for storing a finished lean model is essentially the same as those computer generated files we have created in the first six chapters of this text. Whenever we save something, a file is created. Soon we have hundreds of files running around each containing special symbols, conventions, lean constructions and manufacturing details. The purpose of storing models is related to the concept of layering. Different layers are the contributions of a team of specialists, which include the engineer (data preparation), designer (product design), detailer (image processing) and illustrator (renderman). In some cases one person may often function in several capacities and develop ideas for more than one layer, particularly in a small office. On large projects, the practice is to isolate design functions and activities to a single layer.

1.8 Lean Model Geometry

Geometry for lean model construction is contained in the **AME** pull-down menu described in Chapter 8. In this chapter you'll see how 3D commands like; **ARRAY, FACE, MESH,** and **POLY** are used to **COPY** and **MIRROR** lean images. During construction you will often need to use the **DIVIDE, MEASURE,** and **OFFSET** menu commands. When the geometry construction is complete you can place it in a **BLOCK** as shown in Chapter 8.

Before the geometry can be useful, we must understand what it is, what it is capable of, and the purpose it serves within this book. It is a mixture of geometry (mathematics), 3D computer-aided engineering graphics and digital computing. It is not orthographic projection based upon descriptive geometry which is a mapping of 3D onto a 2D surface, usually a piece of paper or monitor screen. This model geometry, as we use it today, evolved from the transition into the electronic age.

1.9 Vector Lean Models

Vector geometry is the study of graphic statics, model velocity and acceleration analysis. This infers that lean models are structures that withstand stresses. The stresses are either pushing or pulling on each member. A pushing stress is called compression and has a + sign. A pulling stress is called tension and has a - sign. If the stresses are stationary, static vectors are displayed, in motion (dynamics). In either case, the application of statics or dynamics to the solution of model structures is not new and has been in use by engineers for a number of years. The addition of the laptop to solve these is fairly new, however. Like lean geometry (LG) of the last chapter, vector geometry (VG) contains powerful tools for problem solving. When the **AME** equipped laptop is used to automate this problem solution, we call it computer-aided VG. CAVG representation of the forces that act in various members of a lean model structure posses many advantages over manual solution; the primary advantage, beyond presenting a picture of the stresses, is that most problems can be solved with the speed and accuracy of the computer. Combined with the LG skills outlined in Chapter 8, stress may be presented much more accurately than the various members can be sized, since in sizing we must select, from a handbook, members capable of withstanding loads equal to or greater than the design load. Using VG a model designer computes a size and then applies a factor of safety when designing any lean model.

Customarily, problems in statics or dynamics are solved by manual algebraic methods. In this chapter we will assume that you are familiar with one or the other. With no previous background, you will gain little from the study of how to automate it. Before proceeding further; 1) study the screen output shown Chapter 9 which compares LG and VG display methods, 2) notice the items in the pull-down menu, and 3) review the VG terminology listed. Many excellent references exist for studying the terminology. You should consult a reference source if any of the terms are unclear. We begin the study of VG by building directly on the skills learned in the last chapter.

1.10 Wireform Generation for Lean Design

The purpose of chapter 10 is to concentrate on one of the four major types of lean models introduced earlier in this book. A more sophisticated method of describing objects than selecting primitives or BLOCKS as shown in Chapter 6, wireform generation models them in 3-D space. The first of the four methods, and the simplest form of lean modeling is called wireframe. A wireframe model is a skeletal description of a 3-D object. There are no surfaces in a wireframe model. It consists only of lines called wires. Because we use the **AME** package to display this model, the most often used pull-down menu is the model submenu **DISPLAY** as discribed in Chapter 10. This the most efficient method of placing wires within the model. You can set the diameter of each wire, called density. Wireframe lean models can be used in a wide variety of applications to provide a superior method of describing and examining objects as they really exist. Some examples of applications are: 1) Enhancing visualization of 3D objects, 2) Automatically generating orthographic and auxiliary views, 3) Generating a wireframe on which to create surfaces, 4) Visual interference checking, 5) Reducing the need to create prototype models, 6) Approximating volume and mass, 7) Axonometric view construction, 8) Finding intersections of 3-D objects, and 9) Determining the distance between two nonintersecting objects.

1.11 3-D Lean Surface Models

The purpose of chapter 11 is to concentrate on the second of the four major types of lean models introduced earlier in this book. A more sophisticated method of describing objects than wireframes as discussed in Chapter 10. 3D surface models use planes in 3-D space. The second of the four methods, and still a fairly simple form of lean modeling is called 3D surface modeling. A surface model is an infinitely thin shell that corresponds to the shape of the object being modeled. This shell consists of a combination of flat and curved surfaces or adjoining surface elements called patches depending upon the type of surface model used. The main types of surface models are: 1) Extrusions, 2)

Ruled surfaces, 3) 3D faces, 4) 3D meshes, 5) Smoothed meshes, 6) Tabulated surfaces, 7) Revolved surfaces and 8) Edge surfaces.

1.12 Lean Solid Models

The purpose of chapter 12 is to concentrate on the third of the four major types of lean models introduced earlier in this book. A more sophisticated method of describing objects than wireframes or surfaces as shown in Chapters 10 and 11, solid modeling use solids in 3-D space. The third of the four methods is called lean solid modeling. This is a slightly more difficult form of lean modeling. A solid model is a completely filled surface or wireframe corresponding to the shape of the object being modeled. A solid model may consists of a combination of sections called regions or solid composites depending upon the type of lean model created. The main types of solid models are: 1) Extruded regions or solid primitives, 2) Revolved regions or composite solids, 3) Boolean regions or solids like **SOLINT** (intersections), **SOLSUB** (subtractions), and **SOLUNION** (unions), 4) Shaped regions or solids, 5) Smoothed regions or solids and 6) topology regions or solids.

1.13 Lean Section and Profile Models

The purpose of chapter 13 is to concentrate on the fourth major type of lean model introduced earlier in this book. A more detailed method of describing objects than wireframe, surface or solids modeling as shown in Chapters 10, 11 and 12, section and profile models use cross sections of solids in 3-D space called regions. The fourth of the four methods is called regional modeling. This is a slightly more difficult form of lean modeling to visualize in the mind's eye. A region model is a cross section of a wireframe, surface or solid corresponding to the shape of the object being modeled. A solid model may consist of a combination of regions, but a region model can not have a combination of solids. A region is unique to its cross sectional shape.

1.14 Lean Methods of Dimensioning

Filtered dimensioning (XYZ FILTERS) is the process of annotating a lean model to show sizes of model features or the distances or angles between faces, edges and features. Features may include holes, slots, bosses, fillets and rounds, chamfers, counter sinks and bores, and many others. Designers and engineers agree that dimensioning is the most time-consuming part of the preparation of a lean model.

Lean methods simplify the task of dimensioning a lean model as much as possible by automatically calculating distances between selected points and by providing a dialogue box of dimension features for you to begin dimensioning a model. In Chapter 10 we used XYZ filtered database to automatically produce frontal, profile and horizontal (PLAN Command) orthographic working views of a 3-D lean model. Review sections 10.2 and 10.3 if you do not recall this discussion. In this chapter, you will use the same technique to add the dimensions to the orthographic viewports and to the axonometric viewport.

1.15 Dimensioning Lean Models

A lean model is intended to convey   size and shape information regarding every detail of the 3-D object displayed. The model may be anything from a ball bearing 1 mm in diameter to a complete robotic man- ufacturing plant. Without definite specifications expressed by dimensions it is impossible to indicate clearly the design intent that will achieve the successful production of the model. A correctly dimensioned lean model should represent how the model is to be constructed, manufactured, processed or produced. The dimensions should permit ease of production regardless of the production method chosen. A dimensioned model should deal cautiously with the choice of production. The dimensioned display should be flexible, giving the people responsible for production some latitude as to methods.

Although this sounds simple, it is difficult to practice. The model designer has an intimate knowledge of the details and tries to dimension the model so that production personnel do not have to measure or read from the display to find missing dimensions. The dimensioning practice described in this chapter is that recommended by the Special Interest

Group for Graphics (SIGGRAPH), the association for Computing Machinery.

1.16 Lean Tolerancing

In recent years there has been a growing tendency to design lean models to solve highly complex computer-aided engineering problems. The lean model (LM) is a product of the computer-aided design (CAD) process. The CAD display is intended to be used in computer-aided manufacturing (CAM), and if this process is integrated **CADAM** (computer-aided design and manufacturing) is the result. With the ever increasing importance of digital manufacturing processes, the trend in many industries is a gradual shift to an all SI (metric) system of lean dimensioning and tolerancing. The aviation, automotive and electronic industries are now using an all digital system to complete in the world marketplace. The dimensioning system described on this chapter is based upon a LM that is described in millimeters and microns. The micron is used to define the tiny amounts of variation required during manufacturing. The primary advantage that results from using lean dimensioning and tolerancing is the simplification of lean computations.

1.17 Lean Production Models

Lean production modeling discussed in Chapter 17 applications is a process whereby the engineer describes the production plan for the model construction, fabrication or the assembly. If the LM is also part of an automated planning process, it is generally thought of as a production LM. Commonly displayed production models always involve parts fabrication. Parts fabrication includes, but is not limited to the following: 1) Casting, 2) Forging, 3) Extruding, 4) Cold forming (extruding and heading), 5) Stamping, 6) Deep drawing, 7) Spinning, 8) Roll forming, 9) High velocity forming and 10) Machining.

1.18 Lean Gear and Cam Models

The lean design of gears and cams for use as prototype models is often taken for granted because of their apparent simplicity. As shown in the dimensioning chapters, there exists a design file within the software used for this book. It is loaded and used to display simple gear and cam 3-D models. This is useful to place gears or cams on shafts, one shaft turns (driver) and causes the other to turn by means of two bodies in pure rolling contact. Cams are often designed on the basis of this principle. If the speed ratio of the lean model must be exact or a rotary motion must be transferred as a rotation instead of a linear motion, toothed wheels called gears are used in place of a cam. The shafts of gears do not have to be parallel, and are often perpendicular. Special types of gears are displayed for shafts that are not parallel as described in Chapter 18. Here the miter or spiral solid model is used for perpendicular shafts and the worm gear (involute helicoid thread form) makes this nonparallel shaft location possible.

A cam can be displayed as a plate, cylinder or other solid model with a surface of contact so designed as to translate rotary motion to linear motion as shown in Chapter 11. The cam is mounted to the driving shaft, which rotates about a fixed axis (see chapter 9). By the cam rotation, a follower is moved in a definite path. This is one of the applications for this chapter introduced in Chapter 9. The follower may be a point, roller or flat surface.

1.19 Lean Fastener Models

A lean model uses fasteners to show the size and shape information regarding every detail of the 3-D object to be assembled. The completed model is usually an assembly display showing what type of fasteners are supplied from an approved supplier and which are unique to the lean model (LM) and must be produced or manufactured locally. Approved suppliers are members of the Industrial Fasteners Institute (IFI).

Since early 2009 a LM designer can select IFI fasteners directly from the web site *www.industrial-fasteners.org* . In June of 1999 the Fastener Quality Act (FQA) was signed into law. This act greatly reduced the cost of world wide threads and fastener interchangeability used by American industry in its operations.

This chapter along with its discussions is presented as an aid to proper design and display of fasteners or fastener details that can not be down loaded from the IFI. This keeps the emphasis on fastener details for LM design and not fastener displays. The chapter is therefore about: 1) Permanent fasteners (welding, rivets, impact screws, etc.), 2) Assembly fasteners (keys, pins, stud bolts, etc.), and 3) Application of specially designed fasteners in GM displays.

1.20 Lean Manufacturing Models

The essential difference between lean production modeling and regular manufacturing modeling with manufacturing software like; AUTOCAM, FABRICAM, MANUFACTURING EXPERT, NC POLARIS, NC AUTOCODE, SMARTCAM and SURFCAM is in the use to which they are put. A lean production model is used by the engineer to describe the production plan for the model construction, fabrication or the assembly of a LM (lean Model). A LM manufacturing model represents what happens after the production process is decided upon. The production LM is part of an automated planning process, while the manufacturing model is part of the computer-aided machining process called CADAM . Commonly displayed manufacturing models always involve parts formation. Parts formation includes, but is not limited to the following: 1) Sand, grit, shotblasting, 2) Tumbling, Snagging and sawing, 3) Burning, cutting and welding, 4) Shaping, planing, slotting and turning, 5) Milling, drilling and profiling, 6) Hobbing, shaving and broaching, and 7) Grinding, honing, lapping and polishing.

1.21 Chapter Summary

This chapter outlined the remaining chapters of the book and gave you a brief overview of each. When you have studied the many CD stored figures in those chapters you can better understand how a laptop is used along with the menu commands for lean model review. The use of lean models to speed up the design process is absolutely critical as the United States competes in the twenty-first century world wide marketplace.

Begin the study of Lean Modeling for Engineers by inserting the CD that came package with your reference book. Insert this CD in the lap top that you brought to the seminar classroom or are using at home to study this subject. Use **My Computer** from the start menu and click on DVD or CD disk drive and open the CD. You will find twenty chapters, 1-20 shown in small file folders. Click on each that corresponds to the chapter you are studying. There are three types of files, contained in each folder. T represents a Table for that chapter that does not appear in the text. C represents a chapter figure for you to study, there are no figures anywhere in the text. CD stands for an insert image for that chapter. As you read in the text, keep these in mind and refer to your laptop as you study the text.

2

Lean Modeling: The Design Process

Begin the study of lean modeling by inserting the CD that you ordered free of charge from DLR Associates, P.O. Box 11, Sunset, SC 29685. It contains all the diagrams and figures mentioned in following chapters. Select Chapter 2 from the CD and follow along. This book will describe a way to produce acceptable lean model design documentation in the framework of high productivity. In order to do this, certain standards must be set. Standards are meant to offer a common ground for both the development of engineering design communication skills and the interfacing of these skills to computer hardware. The choice of hardware for this text is the laptop computer, it is described in detail in Chapter 3, for now it is important to know that any personal computer can be used with this textbook. The software products described in Chapter 3 are given as choices for use on your laptop. For those readers very familiar with computer hardware and software, skip this chapter.

2.1 Lean Model Design Standards

Before 1990, lean models was done on a catch-as-catch can basis from large mainframe computers. There was no recognized methodology (standards). Since then, and the introduction of the laptop personal computer, standardization efforts have set new guidelines for engineering users. These efforts have fallen along the lines of hardware development known as the laptop (LT) and computer-aided design (CAD) software. While there are many different manufacturers of hardware, each using different operating systems, the software is transportable throughout the equipment manufacturers. Now, different uses of the same software by different manufacturers will have the same capabilities and similar file uses.

The primary goal since 1996 has been to increase the usefulness of such laptop workstations. A lean model for a machine part produced on one laptop workstation should be like another produced by a different manufacturer. In fact the file containing the display instruction should be interchangeable. This is because of the standards developed for portability. The term portability means that a lean model display program used on equipment in this book will be transportable to the equipment used in your work environment. This has come about since the standardization of computer laptop displays. This effort began back in 1998 when several industrial users, programmers, equipment manufacturers and other users formed the GSPC (graphic standards planning committee) under the auspices of SIGGRAPH (Special Interest Group for Graphics) of the ACM (American computer manufacturers). GSC was chartered to develop a methodology and set of functional capabilities for laptop workstations.

2.2 CORE Standard

The intent of the CORE standard was to define an automated framework for most lean model applications including the design of lean models. The functional capabilities of the CORE included the display motions, line weights, lettering, 3-D lean constructions and presentation techniques such as orthographic viewing, a choice of color and display mode for the production of model design displays. The CORE was implemented to various levels depending on whether or not the manufacturer wished to include solid modeling capabilities in addition to wireforms. In other words the planning committee recommended that all manufacturers provide a minimum level (starting package) for simple 3-D wireform models and then add several other options that allowed the user to work modern CAD type displays. The AutoCAD 12.6 release was an original CORE standard software package, it was devoted to 3-D displays, and was hardware dependent. Today's CAD software is a full 3-D solids software package which is hardware independent. Another one of the outcomes from the CORE standards was a common set of hardware terminology for laptop workstations.

2.3 ANSI Standard

Overlapping with the CORE development efforts, the GKS (graphics kernel system) was being developed in Europe. Together the CORE and GKS provided the needed stimulus for the ANSI X3/H3 graphics committee. The best features of the CORE and GKS are now contained in the ANSI X3/H3 standard. The GKS has become the international standard with the most recent modifications made in Japan.

These brief introductions to lean model standards have provided a background to the present use of ANSI standards for all lean modeling design procedures. This text is divided into information, the body of the chapter and summary. The chapters build one upon the other. You are instructed by the body of each chapter as what needs to be learned is presented. It is also part of the body of each chapter to introduce what to do, what is acceptable, and what you should practice or study. The summary will reinforce the presentation of material.

2.4 CAD Methodology for Lean Model Design

This book will show you how to use a laptop, as part of a workstation configuration, and the software packages marketed by various companies. Different packages will be used for the different chapters, for example, a freehand sketching package will be used for the chapter on brainstorming ideas. All of the packages from various companies will show you how the chapters fit together to process instructions. Instructions are important because the laptop workstation can do many different types of tasks. A task is a CAD related job that the engineer wants completed. The person who does the engineering design is capable of probing, questioning, and evaluating facts to reach decisions

- the laptop workstation isn't capable of doing these things. CAD is not twenty-first century push-button design; it is help with the graphic and numeric calculations that are done as part of the lean model design. As a person uses the laptop workstation to assist in the model design, they gain experience and knowledge. They become more proficient in evaluating a situation and deriving a conclusion. Unfortunately, this process also tends to resist new approaches, ideas, and change. CAD users find themselves trying to make things work once they select a software package, and therefore a CAD user often resents changes suggested by others. These changes often cannot be handled by the CAD software purchased. Make sure that your software package has extensions that you can program in a language like Lisp.

With this self-limiting condition, it is not difficult to understand how a user of CAD software will often arrive at a premature conclusion as to what is wanted and how to accomplish it. Although the computer solution is quite rapid and the immediate concept might be satisfactory, it is questionable that the design so conceived would be efficient. In CAD, a sequence of events unfolds in a logical order, forming a design pattern which is common to all projects. Study this sequence shown in Table 2.1 on the CD. The remainder of this chapter will discuss each of these steps.

2.5 Lean Design Start

The model design process begins a logical method by which ideas about design requirements are presented creatively as solutions for problems, which in turn are translated into procedures for transforming raw materials into useful products. This process is divided into 14 sequential steps. The sequence is not a one-pass operation. By necessity, some iteration does occur. Findings at one step may cause the designer to go back several steps, or even force a new start.

To provide a logical design start, the incorporation of a LT symbol at the terminal side of some steps emphasizes the need for human/machine interaction to satisfy all requirements of that step before proceeding to the next design step. If the model design does not survive a step, the feedback path provides a route for the iteration required to correct the deficiency.

2.6 Review Requirements

The second step for model design is to review and understand fully all the requirements. The tendency to jump to conclusions must be avoided. A thorough evaluation of functional needs is essential. It has been said that a model design problem, defined correctly and diagramed through the laptop workstation is virtually solved. Let's take a design problem and list its requirements.

Statement of the problem. The company we work for needs a new class of smart, tough, self-guided industrial handling unit. This robotic unit will work in a warehousing environment picking up and delivering cartons and packages.

Given materials and supplies. This new design unit will use an existing mobile platform called the *navmaster*, shown at the chapter heading. This existing unit has established its reliability in over nineteen years of service. It is basically a robotic sub-turret that will provide the base for a broad array of applications. The existing platform has all-wheel drive and all-wheel steering, offering good agility: forward and backward movement and in-place turning. It has a self-guidance system which allows it to move on and off main routes, take shortcuts, and move in and out of dead-end spurs.

New design elements. We are required to design a specific hardware design called a task turret. A task turret may be a lift, arms, shuttles, carousel or camera. Or it may be a combination of these in order to navigate the warehouse and deliver the packages.

A good engineering model must accomplish its function efficiently and therefore must have a high performance-per-dollar LT input. A CAD laptop workstation is used at the review stage to produce simple ideas. But remember that a LT has no ethics; only a human can evaluate the simple, direct idea. If the design idea does not violate human ethics, it should be considered because the simple idea is the most economical to use, maintain, and manufacture.

2.7 Brainstorm Session

We must now work out a trial solution for our example design in chapter section 2.6. You must be prolific during the brainstorming session. A single idea is a dead end. A designer should repeatedly ask, how else can this be done? You should look at diverse ideas as shown in Table 2.3 on the CD.

As an engineer, you should be careful about being overcritical of an idea at this stage. Innovation is only one step away from a wild idea, and a wild idea can trigger a revolutionary concept. Suppose we try a trigger word technique now? Let's focus on the concept of movement (TO MOVE) in Table 2.4 on the CD.

2.8 Test and Analyze Best Ideas

Once the number of different ideas are sketched and saved, the next step in the design process is to test the best ideas as shown in Table 2.5. The flow chart shows a constant updating of current information, for a design started in the present will not be completed until later, and the actual product will not be manufactured until sometime in the future.

In Table 2.5 three different types of analysis are recommended, functions, schedules and components.. In our example, one analysis may be a computer display of the model unit travel path through the warehouse. The second analyze maybe the planned warehouse path commands listed for testing, and the third may be the signal processing for the robotic unit.

2.9 Refine The Design

The testing and analyses are complete, it is ready to refine the number of options. Select one choice as shown in Table 2.6. In this stage the engineer must concentrate on the shape and form of the design. Several things assist in this:

1) you should be specific and pick one trial solution,
2) standardize parts and eliminate right-left hand parts,

3) joint designs should be finalized and fasteners,

4) part sketches should be placed in temporary files,

5) the numbers of parts involved should be reduced,

6) size and weight may warrant further investigation,

7) modular manufacturer should be investigated, and

8) complete analysis of loads, pressure drops, etc.

2.10 Design Layout And Review

These two stages serve two important functions as illustrated in Tables 2.7 and 2.8; 1) it is the result of the concept stages covered thus far, and 2) it is the beginning of the physical development of the product.

Descriptive notes, not usually found on production displays, are encouraged during these two stages. Notes should specify special manufacturing or assembly instructions. At this time, special drawing displays are made to check patent rights. This is a necessary activity associated with lean modeling.

2.11 Display The Lean Model

Once the design layout is approved, CAD displays are prepared as shown in Table 2.9. The layout sketches are used to produce details of the design. You will learn how to prepare model displays in Chapter 6 of this book. The layout sketches described in detail in Chapter 5 of this book present a concept, while the refinement sketches indicate the material and size of parts. Neither contain all the detailed dimensions and information for manufacture.

Before the model is produced for the CAD display, a final check of the design sketches are made. After computer graphics is finalized and released as a production model, it will require great effort to create a new database or modify an incorrect approach. As discussed in Chapters 14 through 16, the computer generated displays should be clear and concise. they should show all the necessary information such as dimensions with their associated tolerances, materials, and specifications.

2.12 Model Hardware

The quantity of the product to be manufactured, the cost, and complexity of the design are parameters that determine the development. The purpose of model development is to improve the design and therefore the quality of the product. You will learn how to model product design in Chapter 17 of this book.

As shown in Table 2.10, the purpose of hardware modeling is to improve the design and therefore the quality of the product. Some of the steps are used to increase the reliability of the product. One or more of the model stages shown can be used in the development sequence.

1. A pilot model is an experimental item which is assembled to verify the concepts upon which the production item is based. This type of model is often different in size from the manufactured item.
2. A mock-up is a static representation of the manufactured item in which the envelope and shape are depicted in wood or clay.
3. A prototype is a full-size functional working model. It is a dynamic model of the manufactured item.
4. A test article is taken from the first production run as a sampling or testing model. It is an actual manufactured item to be tested for engineering design changes.

2.13 Test The Design Prior to Manufacture

Depending on the product manufactured, different approaches can be used in the testing process. Table 2.11 shows the steps for our example design of the Navmaster Vehicle. The purpose of this stage is to check the manufactured product as modeled, against initial design requirements. You must verify that the design has not wandered off on a tangent, or that a primary objective has not been overlooked.

2.14 Simulate The Manufacture

The engineer in charge of the design follows it through manufacturing. During this process, shop practices are reviewed and modified if necessary. Chapters in this text are sequential and will complete this phase of the lean model design process. The *navmaster* is a real industrial robot manufactured by CYBERMOTION, Roanoke, VA. It is on location working at Glaxo Inc., an R & D pharmaceutical company in the Research Triangle of North Carolina. The CD images for this chapter have been supplied my Mr. Tim Orwig, communications director of CYBERMOTION for the Prentice Hall publication, *CAD for AUTOCAD* users, which I wrote a few years ago.

2.15 Chapter Summary

The purpose of this chapter was to introduce the reader to the lean modeling design process used in industry. The following items were introduced: 1) Typical laptop workstation and its terminology, 2) A 14 step methodology for the engineering design process, 3) Design requirements for example modeling problem, 4) Brainstorm session for ideas, 5) Idea charts for model design sketches, 6) Test and analyze the example modeling design, 7) Refine the design, 8) Design layouts and reviews, 9) Display images, 10) Model the design, 11) Test the design prior to manufacture, and 12) Simulate the manufacture.

3

Lean Model Automation

Professional engineers produce much of their work on lap top computers which plug into a design workstation. I have used, for the last ten years, a Gateway, DELL, HP and Compaq. The Compaq, made by HP is an example of the engineer's lap top. I also use an HP scanner, additional monitor, graphics tablet, printer and file saver. This would be considered a minimum for a lean modeling workstation. These types of networks contain the following: 1) lap top computer processor, 2) large monitor (display screen large enough to handle B size), 3) keyboard, and 4) pointing device.

Several lap tops can be connected to a single large plotter or printer. It would not be cost effective to purchase printers or plotters for individual lap tops in an office. Many engineering offices also connect each lap top to a much larger processor which is used as a system controller. All the models, sketches, notes, bills of materials and the like are stored inside the controller. Usually nothing much is stored in the lap top. This is also cost effective because a single copy of the office software is purchased for the controller, not multiple copies for lap tops.

3.1 Requirements for Lean Model Automation

Lap tops connected to controllers containing custom software are playing an increasing larger part in lean model design efforts. Table 3.1 is an AutoLISP listing of custom designed software written for a central controller. Unfortunately, a person who tries to learn more about lean modeling is limited to the CAD modeling software written in LISP. The programming language source code LISP is stored separately inside the controller. To the experienced user , it is a natural medium for transporting data between simple models and highly complex ones, but to the uninitiated it is a puzzle to be solved. To help solve the puzzle, look at Tables 3.1 and 3.2. With Table 3.1 as a starting point, working lap tops can be categorized by the display software used, see Table 3.2.

It is unlikely that you would purchase all these software packages for your controller. If more than one package is available you should create a logon screen that would similar to:

DLR Associates Lean Modeling Directories

  0. **AutoCAD**
  1. **CADKEY**
  2. **CALab**
  3. **DataCADD**
  4. **GenCAD**
  5. **GenericCADD**
  6. **Home**
  7. **MicroCAD**
  8. **RoboCAD**
  9. **VersaCAD**

:

:

Enter Selection

3.2 Equipment and Materials Necessary

In this section we can select the hardware from Tables 3.3 and 3.4, the operating system from Table 3.5, and the CAD software packages from Table 3.6.   In most cases these components will arrive from the manufacturer in shipping cartons.   Open the system carton first and follow the steps shown on the shipping attachment.   Remove the packing material and dust covers.   Place the system unit on a flat surface large enough to hold all the other pieces of hardware.   Next, unpack the following items:

1. Lap top,
2. graphics tablet,
3. CAD support items (see Table 3.x for tablet template),
4. CAD software packages.

3.3 Workstation Installation

Begin the installation process with the system unit called the controller. This is placed in a central location so that cables can be run, or a wireless router can be used to reach each of the individual laptops located throughout the office.   Each laptop is connected by a USB cable to the controller or by a single wireless router.   The controller is connected to a source of 120 volt power.

You are now ready to run all the equipment tests that come from the manufacturer.   Turn on the power and install the various equipment device drivers, software packages and related items that come on CDs from the suppliers.

3.4 Workstation Software Installation

Table 3.6 lists the software used with this book, when it is all loaded in the system controller it will operate as shown in Table 3.7.   Table 3.5 is a list of operating systems compatible with CAD software listed in Table 3.6.   Only one operating system is used inside the controller and in turn produces two types of software files, DXF and DWG.   These two are translated by AutoConvert, a seamless link to the CAD symbol libraries and **CAD (AME)**. **Sketch**, used in Chapter 5, creates **SKD** files which can be converted to **CAD, Generic CADD, Generic 3D,** as well as **GenCADD,** can all be converted through **AutoConvert**.

AutoConvert comes with support for the CAD products shown in Table 3.6 and other software that imports or exports in DXF (display interchange file), including local programs written within the office, technical illustration programs, scanning software for storage of manual sketches and hand drawn models. You will find the CAD user interface easy to follow, thanks to its menu-driven file support. Most of all, you will find CAD software flexible enough to meet your office needs. Translating an entire directory of files can be done by setting it up through AutoConvert and letting it run unattended while the office is closed.

A CAD package can be used as a complete file editor. In some applications however, other programs must examine displays created by your CAD software or generate displays to be viewed, modified, or plotted. For example, if your have made an complex model using INSERT commands to represent separate piece parts, you can process the display file (.DWG) and produce a bill of materials, or even make energy calculations based on the weight and the number of parts used. Another possibility is to use CAD to describe 3-D model parts that are then sent to the controller for finite element structural analysis. You can compute stresses and displacements and send back information to display the assembled model.

3.5 Graphics Tablet Installation

The graphics tablet template is placed directly on top of the graphics tablet. The template contains four menu areas. They operate in conjunction with the monitor screen menu to provide easy access to all software facilities. The template allows a new user to use a pencil instead of the mouse as a pointing device, a more natural display tool. The template that is packaged with the purchase of the software is designed for an 11 x 11 inch graphics tablet (the most common size), if your tablet is larger then this access the file tablet.dwg. It is supplied on the CAD release shipped with the software as shown at the chapter heading. You can plot the new template at a different scale if your tablet is a different size as shown in book CD 3.1. In order to use the template, you must attach it to your tablet so that it won't shift around during use. Then use the command TABLET CFG typed from the keyboard. This will set the four menu areas plus the display screen pointing area shown early in this chapter. Each menu area is then defined by three points (upper left and bottom corners), while the display screen is defined by the lower left and upper right corners. Tablet CFG will ask you for the number of columns and rows in each menu area, type these in.

The major advantage of the tablet template is that you can point to any software command directly, without typing from the keyboard or using the mouse to click and follow the nested hierarchy imposed by monitor screen menus. For example, the command to draw a line under the DRAW screen menu, but it appears directly on the template under the grouping TOGL/DISP/DRAW section. When LINE in this area of the template is touched, the LINE submenu will automatically appear on the monitor screen. A new user spends a great deal of time finding where things are located.

3.6 Finding Things

The tablet template groups things to find under the following headings: 1) Menu area 2; 3D/ASHADE, DISP/DRAW, 2) Menu area 3; NUMERIC display screen area, and 3) Menu area 4; INQUIRY, TEXT, DIMENSIONS, SETTINGS.

On the template, you will find; OSNAP, EDIT, and UTIL/PLOT. You will need to practice with each of the template items. Some later chapters are devoted to each of these, for example in chapter:3 TEXT, SETTINGS, BLOCKS, OSNAP, EDIT and UTIL, SKETCH, - which permits freehand display to be entered as part of an CAD display (.dwg) file, as well as Sketch software. All of the rest of the tablet menu items are; TOGL, DISP, DRAW, BLOCKS, LAYERS, DIMENSION, EDIT, TEXT, and PLOT.

Each of the remaining chapters is designed to use some of the menu area items and all the chapters will use all the menu area items. A lean model is a file that describes an image made up from the items shown in this chapter. CAD software uses this description to produce the image on the monitor screen or on the plotter paper. In order to create this file you must be aware of the following: 1) World Coordinate System (WCS), 2) User Coordinate System (UCS), 3) Model Display Units, Scaling, Limits, Extents, 4) Insertion, Colors, Linetypes, and 5) Layers.

3.7   Model Display Coordinates

CAD software listed in Table 3.6 uses a fixed Cartesian coordinate system, to locate points within the model. The X axis indicates horizontal distance and Y indicates vertical distance. The origin is where the values of X and Y are zero. CAD calls this the WCS.

The WCS is fixed and can not be altered, however, you can define an arbitrary system called UCS. The origin of the UCS can be anywhere in the WCS and its axes can be turned or tilted any way you choose. UCS allows you to shift the construction plane and simplify 3D location points. For example, the top of a machine part becomes easier if you define a UCS positioned and oriented in relation to the top surface. There is no limit to the number of UCS you can use in a single model.

3.8 Model Display Units, Limits and Extents

CAD software assumes that you will display in a rectangular area. The display limits are the boarders of this rectangle (0,0) at the lower left corner and (10,8) at the upper right hand corner. The display unit is the distance between dots. A represents 1 unit. A unit can correspond to whatever form of measurement you desire. It can be inches, feet, yards, meters, whatever. Therefore, you can draw using real world units and eliminate the possibility of scaling error. When you have finished the model, you can plot it at any scale you like.

The limits of a model might be 1000 yards for a site plan, or metric units for a part. Just as the limits specify the potential size of your display, the extents show how much of the area within the display limits contains model information. Imagine a sheet of paper is placed over a large lean model display, move the paper around on top of the display.

The display is the limit, while what is under the paper is the extent of the viewing of this model display.

3.9 Model Display Entities

The process of model display consists of the placement of entities selected from the tablet template and placed at coordinate locations. You may select any of the model display entities listed in Table 3.9.

3.9 Model Display Insertion

You can insert an existing model stored in the system controller as a **BLOCK** and merge it into the model that you are currently creating. In this case a simple 3-D shape was constructed as a display part and stored as a regular (.dwg) file, and then inserted as many times as needed. Using this technique, you can construct a library of typical modeling parts that are used often in the office.

3.10 Colors and Linetypes

Different colors are sometimes used to indicate cold or hot surfaces on a lean model. CAD provides 255 different colors, more than you will need for even colored technical illustrations. Linetypes, however, are very important as a user you must be able to draw the following linetypes:

1. Object _____
2. Construction _____
3. Section _____
4. Center _____
5. Phantom _____
6. Hidden _____

3. 11 Layers

One of the most useful and most often used model technique is **LAYERS**. You can assign various portions of an lean model to different layers, and you can define as many layers as you like. You can think of layers as transparent sheets with information on each separate sheet. You can view (turn on) as many layers as you wish. In this book we see the layering concepts used in Chapter headings. Layer one contains a completed lean model that was inserted from a file (PATCH.DWG), layer 2 has the bearing and shafts, while layer 3 contains a bushing inserted from file (TUTOR.DWG). The text Intro to can be placed on a separate layer. Stressed parts on another layer and so on. You can freeze selected layers of your model. A frozen layer is excluded when the model is regenerated, thereby speeding up the gen time. When you like the frozen layer can be thawed for storage or plotting.

3.12 Finding Model Files

Model files are stored after they are created at the laptop workstation. The central controller stores these file under the following directories.

1. **cad - displays created under your software,**
2. **sketch - displays created under SKETCH,**
3. **generic - displays created under genericCADD,**
4. **gcad - displays created under generic 3D,**
5. **gencad - displays created under gencadd.**

To find a particular file you must request the directory where it is stored. Software companies will ship several CD models with your purchased software. Let us begin with the cad directory. You may select this directory by typing acad from the keyboard. This will cause the following screen or similar menu to appear:

Main Menu

0. **Exit CAD**
1. **Begin a NEW display**

2. **Edit an EXISTING display**
3. **Plot a display**
4. **Print a display**
5. **Configure CAD**
6. **File utilities**
7. **Compile shape/font description file**
8. **Convert old display file**
  :
  :

Enter selection

Select 2 from the menu. The controller asks for a display name, type TUTOR from the keyboard and the stored display TUTOR.DWG is displayed for you.

3.13 Editing an Existing Model

We are going to try several editing techniques with this display. You may type commands from the keyboard, select from the pull-down menus at the top of the display screen, or point to the various commands and functions on the graphics tablet template. Select PAN from the tablet under the EDIT section in menu area 4. We need to check and see what the limits of the display are, we can see the extents on the monitor. We have PAN right or left in any of the CD files. This allows us to see the limits of the display file TRACTOR or TUTOR . It would appear that the file TRACTOR is a 3D display so we can practice our use of menu area 2, 3D/ASHADE. You will find HIDE at the upper left hand corner of this menu. Select it and compare your monitor screen with what you saw before. Before you leave this menu area you can pick the second menu item and it will ask you for the VPOINT, type in (0,0,1) from the keyboard. Or an easier way is to press PLAN in the next set of menu items. You can produce front elevation views. When you have finished playing with menu area 2 you can return to the display by entering VPOINT(1,1,1).

You may try all of the menu items in EDIT, if you want to change the display by deleting areas like the tires, wheels, grader blade and so on, use ERASE. Compare your display. When you are tired of working with this CD display you select another from the following:

1. blocks.dwg, bulb.dwg, dim.dwg,
2. disc.dwg, hatching.dwg, hub.dwg,
3. lamp.dwg, site.dwg, switch.dwg,
4. kitchen.dwg, pins.dwg, or asetut.dwg.

Be sure you try all of the menu items.

3.14 Begin a New Model Display

Select main DLR menu item 1. LT responds.

Enter NAME of display:

Type in LOGO. Do not add (.dwg), the controller will do this for you. Let's create a business logo that can be used over and over again with INSERT. It should fit at the bottom of a size A sheet (9x12 inches) or at the right side of a size B sheet (12x18). You may use the example shown on the CD or design your own.

Begin the display process at the left side of the title strip, use ARC to display the left hand side. To draw an ARC, you need to select a start point. Here are the ways to specify start points:

1. pick a point on the screen with your pointing device,
2. pick a point using ORTHO, SNAP, and GRID,
3. use OSNAP (object snap) to specify point on existing geometry,
4. pick a point by typing coordinate point locations on the keyboard.

The most time consuming and least effective is #4 followed by #1. The most effective use of time is #2 where no previous geometry exits and #3 where you are connecting to existing display images. We will use method 2 to display the first and second arc, then 3 to connect a line between the arcs. Move over to the start point of the arcs on the right side of the logo and the repeat the process. Remember to use SNAP, so the arcs will line up with those on the left side of your display screen. Now connect the two arcs at each side of the display with LINE.

Because this title logo will be used as an INSERT, we will need to use TEXT to enter the text shown in our example above. You are now ready to SAVE the competed display for use later.

3.15 Plot a Lean Model Display

Main DLR menu item 3, if selected, produces a plotted display on a pen plotter. Because you are at the main menu, you can plot any existing display. You may select the logo you just completed or you may any select any of the CD displays used in chapter section 3.13.

   You may plot a display while editing, instead of from the main menu, just select the PLOT: command from the display screen or from the EDIT menu from the tablet. Usually, a REDRAW command while editing is better than a pen plotted hard copy.

3.16 Print a Model Display

Main DLR menu item 4, if selected, produces a printed copy. Because you are at the main menu, you can print any existing display file. You may elect to print while editing or creating a new display by using PRPLOT.

3.17 Configure AME

Main DLR menu item 5 is used when you install the CAD software purchased with your laptop workstation. Before using cad from the controller, CAD along with several other graphics packages were installed. The Configure CAD function allowed the installer to select the drivers for your laptop workstation hardware (printer, tablet, mouse, and others). This was done for you by someone else, you may use it occasionally to change devices or defaults.

3.18 File Utilities

Main DLR menu item 6 is used to pass control to the controller's disk file utility submenu, from which you can list the contents of a disk, delete display files, change the name of a file, or copy a file.

3.19 Compile Shape/Font File

Main DLR menu item 7 converts SHAPE descriptions into a form usable by the CAD's display editor. You will need this menu item when creating or modifying a SHAPE or FONT file; see Chapter 4.

3.20 Convert Old Model Display Files

Main DLR menu item 8 uses AutoConvert. From time to time during the design of a lean model we use different directories within the DLR software. A sketch created with **SKETCH** is stored in the sketch directory as name.skd. Therefore if we **INSERT** name.skd into a file. dwg, it must be converted from skd to dwg file format. Keep in mind that once converted a file remains converted. Therefore, remember to make copies of files and convert the copies if you want name.skd to remain in the sketch directory.

3.21 Changing Directories

Changing from one directory to another is different than transporting one file into another. In order to leave cad, select main menu item 0. Remember you can then select from; cad, sketch, generic, gcad or gencad. Book CD, C 3.2 through 3.5 represent different directories. C 3.2 is the SKETCH directory. C 3.3 is the GENERIC directory. C 3.4 is the GCAD directory and C 3.5 is the GENCAD directory.

3.22 Directory Format Styles

It is easy to see what directory you are working in by the format style shown on your laptop. We have been working in a DLR directory. The format shown in Tables 3.1 through 9 indicate this. Let's change the directory to sketch. You will notice that there are now pull-down menus at the top of the screen. These are Draw, Change, View, Assist, Settings, Measure, File, % of file space used, and time sketch was created. There is no side menu with Sketch. Sketch can not be activated from the graphics tablet menus.

3.23 Chapter Summary

The automation of the engineering office to produce lean models was discussed beginning with the requirements in hardware and software. In order to use this book, you must be familiar with the various software packages. If you are familiar with CAD software, you may skip this chapter and enter *www.e-model.net* or *www.mkp.com/books_catalog/area* on the internet. If you are not familiar with CAD packages, then you should look at the various chapter headings, study the software and hardware listings, note the operating systems used, then install the software and hardware, find out how to find things, try CD displays stored in the software and practice display. Familiarize yourself with display insertion, colors and linetypes, layers, display files, and cad main menu items.

4

Notation-Surface Description

This chapter has a unique approach to lean model notation and surface description, stylized solely for the production of lean model surfaces. The styles of lettering and surface notation provided are shown in Table 4.1, one of these may be selected as the default office acceptable lettering that appears on lean models. In addition to the default lettering style, special shape fonts can also be formed. After a special font style is made, however, most software has the ability to generate this symbol. The purpose of this chapter then is to develop a system which will consist of the workstation requirements in both hardware (scanner) and software instructions **ATEXT.shp**. Then we will describe how to use the equipment, and how to code the surface description fonts for storage in the system controller. Samples of text styles look like:

Text Styles Available in CAD

SAMPLE NOTATION STYLE CHANGED TO ZURICH

SAMPLE OF LEAN NOTATION STYLE CHANGED TO VRINDA

SAMPLE NOTATION STYLE CHANGED TO VERDANA

SAMPLE OF LEAN NOTATION STYLE CHANGED TO TREBUCKET

SAMPLE OF LEAN NOTATION STYLE CHANGED TO TEMPUS

SAMPLE OF LEAN NOTATION STYLE CHANGED TO SWISS
SAMPLE NOTATION STYLE CHANGE TO SEGEO
Sample notation Style change Segeo lower case

4.1 Use of Existing Lettering Fonts

Before we design our own surface notations **ATEXT**, much can be learned from the use of notation commands: **TEXT, TEXT C, TEXT M, TEXT R, TEXT A, TEXT F, TEXT S, DTEXT, STYLE, LOAD** and **SHAPE**. You are allowed to modify existing styles and this may be the way in which you decide upon the common default type surface specification and lettering.

4.2 TEXT Command

When this command is used you can control both the text font (shape) and style (appearance). We add notation or surface symbols by selecting this command. The notation can be displayed with a variety of patterns called fonts, can be can stretched, compressed, obliqued, mirrored or displayed in a vertical column by applying a **STYLE** to the font. Each line of lettering can be rotated and justified to fit the model space or screen requirements. The **TEXT** command appears below the command line on the monitor as:

Command: TEXT
Start / Align/Center/Fit/Middle/Right/Style:

at this point you must select one of the options given by typing S, A, C, F, M, R, or S. These actions are shown as:

TEXT COMMAND OPTIONS

Keystroke	*Definition*
Start point	*Left justifies the text baseline*
A	*Prompts for 2 endpoints*
C	*Centers text line at that point*

F	*Sizes the text to fit the space*
M	*Centers text*
R	*Right justifies the text*
S	*Asks for new style of text*
Null reply	*Advances immediately*

If we do not respond (null reply), the default options are selected and the screen looks like:

Command: TEXT
Start point or Align/Center/Fit/Middle/Right/Style:
Height <0.20>:
Rotation angle <0>:
Text:

If you leave the height null, .2 inch is selected for you and rotation is 0. You must enter the line of lettering, otherwise no lettering is visible. For multiple lines of lettering as shown in the CD figures, the software assumes that you want to place another line of text below the previous line, at the same angle, on the same layer, with the same style, color, and height. When you hit the return at the end of the first line, the command line responds with:

Command: (hit return)
TEXT Start point or A/C/F/M/R/Style: (hit return)
Text: (type second line here)

4.3 Selecting a Text Style

AME software supplies three types of lettering styles: STANDARD, STRETCH and FANCY. The text style determines the appearance of the text characters. The text font styles supplied by AME are shown in Table 4.1. The special symbols we would like to have available, requires us to create a load file and shape file. We can modify the existing style fonts. Either ROMAND or SCRIPTC comes as close to the desired style as we can get with little effort. If you want to change the current text style for your lettering, respond by entering S:

Command: TEXT
Start point or A/C/F/M/R/S: S
Style name <standard>:

You are now ready to enter the desired style, FANCY for example. Using this method you must chose one of the existing text styles.

4.4 Multi-font Notation

A multi-font form of notation uses multi-font commands, the first is given a null response to set up the second line of lettering. It has two lettering fonts, ROMANS and ATEXT. Atext symbol font can also be used in a single display. For example, the first start point is given , the command is given as **ATEXT** and the second start point is given, with the same command. The special symbol for surface finish is then shown on the LT.

It is often desired to underline lean model lettering this can be done with the addition of a control code character. To use this command add %% code from Table 4.2 before and after the text line, the command line should look like:

Command: TEXT
Start point or A/C/F/M/R/Style: 1.6,4.9
Height: .5
Rotation angle: 0
Text: %%uWIREFORM MODEL%%u

The screen results looks like this:

WIREFORM   MODEL

The special character code %%nnn can be any number between 000 and 127 as:

CONTROL CODES AND SPECIAL CHARACTERS

%% Code	*Definition*
%%O	*Overline mode on/off*
%%u	*Underline mode on/off*
%%d	*Display degrees symbol*
%%p	*Display plus/minus symbol*
%%c	*Display on center symbol*
%%%	*Display a single % sign*
%%nnn	*Display ASCII character*

4.5 Dynamic Text

The lettering command **DTEXT** is used more often than the **TEXT** command because this allows you to see the lettering on the screen as you enter it. You can perform basic editing using the backspace key, and enter multiple lines of text in one command. The sequence of prompts is the same as **TEXT** except that the Text: prompt is repeated. When Text: appears below the command line, **DTEXT** draws a cursor box on the screen at the text start point the size you entered for the height. You can cancel the **DTEXT** line of lettering by pressing the CTRL key and the C key. This is much quicker than using a backspace or **ERASE** command. You can enter the %% codes from Table 4.2 using DTEXT, for example if you entered %%126 an equal sign would appear in the line of lettering. It is often desirable to stretch or compress the lettering, apply a slant to them, or have them placed in a vertical column. This is done by the **STYLE** command.

4.6 DTEXT Style Command

You use the **STYLE** command to create and modify **DTEXT** definitions, to list the current styles, or to set a particular style as the current style. when you begin a new drawing, a standard text style is automatically created. It is used for all **DTEXT** items until you create another style and request its use. You create a new style by selecting menu item 7 from the main menu just before you begin a new drawing.

The display screen looks like this below the command line:

Command: STYLE DTEXT style name <standard>:
Font file <txt>:
Height <.2>:
Width factor <1.>:
Obliquing angle <0.>:
Vertical <Y/N>:

The default information appears between the < > marks, at this point we can change the style name from standard to any of those shown in Table 4.. The font file can be changed from TXT to any of those listed. The height of the lettering can be anything from .1 to the height of the display screen. If you set the height to zero, then each text entity can be set individually. The width factor can be set above or below the default value. One is a normal width for each letter, while 2 would stretch each letter double its normal width and .5 would decrease the normal width by half. The obliquing angle produces inclined lettering. The default value is no inclination, 15 produces 15 degrees of inclination. A -15 produces a backhanded inclination of 15 degrees. And finally, the vertical can be turned off by N and turned on by Y.

4.7 Shape Commands

Imagine that the text character is displayed like a scoreboard character. The DTEXT fonts use this scoreboard matrix and connect the dots with lines, arcs and circles. Once connected you can then display this matrix from a LOAD file which contains its pattern. You use the SHAPE command to put patterns from the LOAD file into your drawing. Shapes are defined by special files with a file type of (.shx). In the case of the letter A, it was described by the file description *065, because the ASCII code from Table 4.3 is number 65. Shape files that begin with *065 through *090 are capital letters and *097 through *122 are the lower case letters. The origin for each shape is the lower left hand corner of the matrix. The file for the capital A contains seven pieces of information and would look like this:

*065, 7, ALETTER
MOVE TO ORIGIN (002)
DRAW VERT LINE (001)
DRAW TOP ARC (008)
DRAW DOWN VERT(001)
MOVE UP 3 (001)
DRAW HORIZ (001)
END (000)

Additional shape codes are:

Codes	Definition
000	End of shape definition
001	Activate display image
002	De-activate display image
003	Divide vector lengths by next byte
004	Multiply vector lengths by next byte
005	Push current location onto stack
006	Pop current location from stack
007	Display shape number given by next byte
008	X-Y displacement given by next two bytes
009	Multiple X-Y displacements term by (0,0)
00A	Octant arc defined by next two bytes
00B	Fractional arc defined by next two bytes
00C	Arc defined by next five bytes
00D	Multiple bulge-specified arcs
00E	Process next command

4.8 Load Command

In order to better understand the relationship between shape files and LOAD commands, suppose we enter the **LOAD** command as:

Command: LOAD
Name of shape file to load (or ?): ?
Loaded shape files:
>   **pc**
>   **es**
>   **hardware**

notice that we responded with a ?, which gave us the shape files currently in the controller. If you list these files you can better understand how to write your own shapes as we did for ATEXT shown as:

File: pc
FEEDTHRU   DIP8 DIP14 DIP16 DIP24 DIP40

4.9 Using the LT and Scanner

Not all associated symbols used with lettering need to be defined and stored in **SHAPE** files. A hardware approach using a scanner can create **BLOCKS** which provide the primary means of defining and using a library of model symbols and parts. All of the associated symbols can be scanned, stored in blocks and inserted.

You can open the scanner driver using your software. Depending upon the scanner manufacturer, you may need to select the driver shipped with the scanner and enter it into the controller. For details, please refer to the instructions that came with your scanner. Table 4.5 lists the main screen features and their definitions are:

Scanner Main Screen Features

Main Screen Item	*Definition*

Menu Bar	*Adjust settings and display driver*
Comb Boxes	*Specify different scan settings according to*
original	
Preview Area	*Preview the image in order see*
Tool Bar	*Access several tools in order to adjust the*
scanned image	
Status Area	*Display information of current image*
Command Button	*Mange scan jobs and control scan actions*

4.10 Scanning Associated Lettering Symbols

Place the existing original lettering or symbols face down on the scanner glass plate. Note the direction of the original so that you will not scan the image in the wrong direction. Close the scanner lid. Open your receiving software (this produces the scanner main menu). From the menu select Reflective in the original combo box, then click Preview. A preview image will appear in the preview area. You may adjust the scan area in the preview area.

Use the options in the Combo Boxes to specify the resolution, scale and the others that will apply to your original. Use the options in the Tool Bar to adjust the image. If you need to add another scan area to the original, push the Duplicate button in the Job List to add this. Then repeat the adjustments for this new image. Now press the Scan button.

4.11 Scanning Surface Specification Formats

Not all existing originals are pencil or ink on reflective paper. Tracing paper, drawing film, colored slides and filmstrip all contain previously prepared and stored lean symbol originals. You may scan all of these in the following ways: 1) Place your original in an appropriate film frame, then place on the glass plate of the scanner, 2) Make sure the calibration area is facing the right direction (towards the front of the scanner), 3) Do not block the calibration area, or the scanner won't work, and 4) Use the Combo boxes on the right side of the scanner dialogue box. Following the methods just described, you can scan all of these materials by selecting the corresponding commands from the combo boxes or the main menu.

4.12 Selecting Associated Notation Items

You will want to select only those items that can be placed in a notation use library, a few examples might be: 1) General notes format, 2) Surface finish symbol, 3) Surface descriptions, and 4) Recommended surface waviness height values. Each of these can be created with the BLOCK command and inserted into a lean model display.

4.13 Chapter Summary

The purpose of this chapter was to develop a better understanding of how lean notation is handled within software and how this type of lettering can be modified for use in an engineering design office. The chapter introduced the common **TEXT** commands (**TEXT, DTEXT, STYLE, LOAD** and **SHAPE**). An entirely new font was developed called **ATEXT**. The use of a workstation scanner was covered with the building of lettering blocks to be used over and over again during the design process.

5

Lean Model Sketching

Much of the lean model construction in the first four Chapters was in
SKETCH format. **SKETCH** is not a freehand sketch. It is an automated
format for using a simple quick method for capturing ideas. It would
be impractical to design anything by starting with a CAD generated
lean model before preliminary concepts and ideas are discussed and
approved. It would also be even more difficult to design a model from
purely verbal descriptions. The research and planning for a lean model
include idea sketches for communication, preliminary idea diagrams,
notes, dimensions, and specifications. All these will be needed later
in the final preparation of a model. Freehand model sketching is
introduced in this chapter directly after automated notation so that
the quality of the lettering, sketching and completed model will be
compatible.

When you finish this chapter, you should be able to model curved
as well as straight line parts with freehand circular holes, rounds, fillets
and curved surfaces. Your sketches will be similar to those shown at
the chapter heading. This chapter contains sketching instructions to
create a working database. This database maybe obtained by the use
of a graphics tablet as shown in Chapter 3. In this chapter we use the

graphics tablet and pointing device (pen) just like a lead pencil and engineering cross-section paper. Throughout the material in this chapter the concept of laptop (LT) together with our chosen software is stressed, beginning with the graphics tablet and freehand sketching commands and moves on to the other software packages. The chapter ends with the use of the LT scanner and the building of a sketching library stored as BLOCKS.

5.1 Types of Sketching Models

Manually drawn freehand sketches are extremely fast to produce and are very cost effective for model communication. After they are completed, however, don't just stick them in the project file. Take them to the scanner and enter them as drawing files with the file extension .SKD. These freehand sketches can now be stored inside **SKETCH**, or run through **AUTOCONVERT** to make .DWG files. Better yet switch from paper and pencil and complete your sketches on the graphics tablet and they are stored directly as .DWG drawing data base.

Like model annotation, model sketching is one of the fundamental forms of presenting technical data. In fact, if only character and shape generation plus model sketching are used, lean modeling could be taught and learned by most. Model sketching is valuable to every one, more so now than ever before. The scientific age in which we live requires many abilities in order to communicate with co-workers. Accuracy in observation is developed by modeling. There are those who learn modeling for no other reason than to see what they are about to design. Many engineers look but do not visualize an object such as a machine part.

When describing a complicated object, word descriptions are usually very difficult. A sketched model would eliminate the need for a lengthy and involved word description as is the case of most computer programs. Only a short supplemental word grouping called a note is required with a sketched model. An interesting experiment is to sketch a simple object, then describe the object in as many words as needed to fully describe it. Now ask a friend to sketch the object from the written description only. Compare the two sketches.

The design of any object starts with an idea. The designer sees the idea in the mind's eye and uses freehand sketching to aid in the thinking process. The accuracy of the presentation of the ideas to others is not as important as the communication of the idea. Freehand models may be employed as an aid in visualizing and organizing design problems as well as ideas. Most engineers develop the ability to think with a pencil. Freehand models are made of such things as force vectors (Chapter 9), parts and pieces (Chapter 6), connections (Chapter 8), and trial solutions (Chapter 1)

5.2 Using the Graphics Tablet

A graphics tablet as introduced in Chapter 3 was used as a large menu selection.   In this chapter we can use of the entire tablet surface for sketching if we want. This can be awkward due to large distances the tablet's cursor (pointing pen) must be moved to effect a small echo display on the monitor. Therefore, you should use the tablet as configured for Chapter 3.   Remember we used the TABLET CFG command to do this.

Because we are still in the DLR directory we can use the SKETCH command to begin our freehand sketching. This command is useful because as the tablet pointer is moved the data is entered into the .dwg file on the fly. The file is composed of points and lines only and therefore these types of files are much larger than normal .dwg files. You are allowed to edit shapes before you store them however. The command looks like:

Command: SKETCH
Record increment <0.1>:Sketch. Pen eXit Quit   Record   Erase
Connect o

You must enter the increment (distance length of each line segment). This establishes the resolution or accuracy of the sketch. If you want to leave the increment at 0.1, hit return and the sketch menu line appears. These stand for P: pen up/down, X: store sketch and exit freehand sketching mode, Q: discard lines and exit, R: record (store), E: erase backward by pointer movement, C: connect line to line, o connect line to point.

You begin a freehand sketch by pressing the P key. Touch the tablet with the pointer and see if the display monitor shows a line. If not, the pen is up, hit P key again, now notice that a line is displayed. It will take some practice to get the hang of this. Try something simple like the sketches at the chapter heading.

You can practice the use of the other keyboard menu items, try E and move backward along the line just sketched. Next R an object, C the object, o the sketch, and Q the completed sketch. You can X the entire screen and move to an intermediate sketch.

Machine tools are shown in a presentation sketch. It is important to remember that all of the LT commands are available to use with the SKETCH command because you are still inside the LT directory. Let's add the outline of a sample sketch with LINE commands. You may want to review the LINE command from Chapter 3, use of the tablet menu. And finally let's learn a new command, RENDER. It looks like:

Command: Render

and causes a dialogue box to appear on the screen. The quick render options dialogue box let's you control the following:

1. Output mode - color or black and white.
2. Back face options - discard or normal.
3. Separation - Black and White or red, green, blue.

The completed rendered sketch is simple. If the dialogue box does not appear you can bring it up with the command RPREF
The rendering preferences dialog box let's you control the following:

1. Rendering type - determines speed and quality of rendering.
2. Rendering options - smooth shading or merge.
3. Select query - determines whether you select a portion of the sketch or whether the whole screen is rendered.

4. Destination - choose output device, screen, printer or plotter.
5. Color map usage.
6. Settings - controls the light, and turns LT Ashade on.
7. Information.

5.3 Freehand Sketching Techniques

The purpose of freehand sketching in the LT directory is to make a user approaching the study of lean models a highly productive individual because of the additional options available in acad. A user qualified in the use of a graphics workstation, scanner, and graphics tablet can assist in simple research procedures and final lean models. The user must have 3-D ability - the computer only stores and repeats the designer's sketch in various formats in a very fast time frame. These elements are basic to the speculative nature of engineering office work and attracting new product designs.

5.4 Communication Using Lean Models

Freehand sketching is the best means of capturing, developing, and storing ideas in the planning and meeting stage of a lean model design project. The principle and procedures are simple, and only a few basic techniques are needed. Following the procedures from the last section of this chapter, a user may prepare a LT to accept, modify, store, and reproduce sketches that can be shown to a co-worker or machine tool purchase customer.

Sample sketches shown in C5.5 through C5.11, on your book CD, represent a design office series beginning with the newly proposed machine tool displayed as a lean wireform (C5.5), the customer is shown floor locations A-B (C 5.6). Next a solid model can be presented to demonstrate machine tool motions (C5.7). The customer can even see a sketch of associated machine tool operations such as part loading (C5.8). The customer may even study the affects of machine tool movements on shop pathways (C5.9), or what effect a model shape change has between angular or linear velocity (C5.10). And finally you see how the machine tool operational forces affect the angular and linear acceleration of the lean model (C5.11).

If the customer wishes to change anything, the sketches are stored in .dwg files on the CD and can be easily modified as the customer watches. The best parts of the sketches can be saved, and the new approved sketches are produced without re-sketching by hand. Good freehand customer sketches look freehand. Do not try to make a model sketch look like some mechanical instrument drawing. At the start of customer communication, too much precision in making models wastes time. The detailed models can be produced after the customer's approval, as shown in Chapters 6 and 7. The following are a few pointers to LT directory freehand users: 1) Use free arm motion with the graphics tablet pointer, 2) Avoid a cramped finger stroke, 3) Do not make all corners precise, but let them cross, 4) Make sketches fit the screen as shown in this chapter, 5) Multiple (polylines) are often more effective than single lines, 6) Be careful of proportions; don't worry about actual dimensions of the product, and 7) Try to avoid small details in model sketches; the overall effect is more important.

5.5 Model Construction

Model construction involves lean positions. A model should be constructed so that it appears in its natural position. The data collection system used in all the CD figures in this chapter assumes a traditional rectangular coordinate system where X and Y are mutually perpendicular and Z intersects them to form three mutually perpendicular reference axes. To obtain the most descriptive position form which to view the final model, the observer should walk around the model noting all features. The position that best describes the model should be selected. You will note in LT sample sketch (C5.1), that the traditional viewing axes have been by this walk-around technique.

Most sketching systems have selective viewing as shown in (C5.12). The viewing position may be changed by entering a YES or NO and the amount of axis orientation desired under PHI and THETA. Sample sketch, C5.12, represents a simple wireform lean model with the following:

A PHI=0.   THETA=0.
B PHI=45.   THETA=45.
C PHI=180.   THETA=180.
D PHI= 270.   THETA=270.
E PHI=45.   THETA=180.
F PHI=360.   THETA=360.

There are an infinite number of positions in which a wireform model may be viewed after the data are determined. In C5.13, the data are listed along with the viewing selection for that wireform.

5.6 Preliminary Sketches

Preliminary sketches all require construction weight lines. These are placed quickly upon the screen and are used to sketch circles, parallel lines on surfaces, edges and faces. Begin your practice of using construction lines as shown in Chapter 3 and in this chapter heading. Next move to the right and layout the rhombus. Move upward on the screen and practice the construction of a circle as shown in the middle of the screen. Next move upward again and practice the construction of an ellipse as shown at the top of the screen. Circles can appear as ellipses by using the viewing position described in chapter 3, but sometimes cylinders are cutoff at angles and produce elliptical lean model shapes.

5.7 Construction Layouts VS Data Orientation

C5.13, was an example of PHI/THETA view positions for raw model data. Sample C5.14, was an example of model construction using light layout guidelines. The application of layout construction is shown in C15.15. Any lean model is made up of lean shapes and forms. It is essential that users learn how to model the more common building blocks so that they may be constructed on a graphics tablet and stored in computer memory for later use.

5.8 Changing LT Directories to Find Sᴋᴇᴛᴄʜ

We are now ready to return to the main menu for the DLR Associates system controller. Up to now we have been working in the LT directory, we will need to select the sketch directory in order to find **Sᴋᴇᴛᴄʜ**. After the building blocks have been entered into the LT directory, you may begin to decide how the assembled lean model will appear. A sketch output of an assembled lean model is shown inside the LT directory for your reference. We notice that the screen format is different because there are no side menu items to chose and the tablet menu does not work in this directory.

Before entering a sketch, you should be familiar with the use of **Sᴋᴇᴛᴄʜ** software. The remainder of this chapter will familiarize you with this software package.

5.9 Pull-down Menus and Quick Sketches

At this point you are ready to make some sketches of lean models. To do this, it will be necessary to determine what the items under the Draw pull-down menu do in Sketch. We need to cover those that are new to us. For example:

1) Arc - functions the same as LT directory, 2) Box - a 2D version of the 3D command, 3) Circle - same, 4) Curve - used to sketch continuous lines, 5) Ellipse - same, 6) Line - same, 7) Part - used in sketch same as BLOCK, 8) Paternfill - same, 9) Point - same, 10) Polyline - same, 11) Quick Text - same as DTEXT and
12) TextEditor - provides a dialogue box.

5.10 Editing Sketches

When you edit sketched models, the Change pull-down menu is used to: 1) Undo - same as E menu item, 2) Redo - undo an undo (change your mind), 3) Erase - a more selective form of E menu, 4) Group - consider more than one image, 5) Ungroup - break apart last grouping, 6) Move - let's you move framing sections, 7) Copy - let's you repeat a framing element, 8) Stretch - let's you change X or Y scale factor, 9)

Property - property dialogue box, 10) Rotate - same, 11) Scale - same, 12) Mirror - same and 13) Break - same.

Many types of lean models can be constructed using the SKETCH software; some to the more common types for are: 1) Wireforms - see Chapter 10, 2) three-D Surface - see Chapter 11, 3) Solid - see Chapter 12 and 4) Section and Profile - see Chapter 13. Many different combinations of these models can be used for the quick solution of a design problem. The SKETCH user must combine them in a way that will suit a co-worker or the final product customer.

5.11 Using the VIEW menu

The view menu contains those items that let us look at a sketch the same way acad did. If you understood how ZOOM, PAN and Redraw worked in LT then you can use everything in this pull-down menu.

5.12 The ASSIST Menu

All of the items that can be toggled on/off in the LT directory are provided in this menu. If you understood ORTHO, GRID, SNAP and ATTACH, then you will use these items in the same way. A grid is either on or off, it is sometimes convenient to sketch with the aid of a grid, just as snapping all lines to the grid is useful. Orthographic sketches are rarely used, but come in handy with cabinet drawings. To be able to attach one sketched item to another is very helpful and time saving.

5.13 The SETTINGS Menu

All of the items have dialogue boxes and require you to enter settings data. Notice that Attach, Curve, Grid, Property, Snap and Text are all repeated in this menu, when you use any of these you may have to change some of the settings before they are the desired sizes, shapes, colors or linetypes.

5.14 The MEASURE Menu

The items are used to dimension sketches. Usually only construction detail sketches require dimensions like Distance, Angle or Area while lean models require Point and Bearing dimensions. The Align, Horiz, and Vert are the same, while the show properties is used to check settings to see it you need that menu pulled down.

5.15 The FILE Menu

This menu is commonly called the house keeping menu. Fourteen file manipulations may be used during the sketching session. You may begin the process of learning how these work by pulling down the menu. To select a file technique, move the pointer to it and press the pointer button. You may select these now one at a time and observe the screen display, notice that these operate just like the LT directory, except they are all in one convenient place.

1. New - submenu item removes everything form the working screen and creates a new file space; if you were working inside another file or sketching session, a dialogue box appears. Select either Save or Discard with the pointer to create a new file space. The Cancel box returns you to the current sketch file.
2. Open - causes a dialogue box to appear. So items can be entered to open a file. Two options exist at this time; (1) stay in directory A and use the file called PARTS, or (2) change the directory depending upon where the sketches are to be stored. To change directories, place the pointer inside the A: box and type the new directory name.
3. Save - sketches are placed in a directory by the save submenu item.
4. Save as - works and looks like Save and can be used to save a different version of a saved sketch.
5. Make DXF - to store a data exchange file (DXF) to be used by other software packages.
6. Pen info - choose from a dialogue box:
7. Plot area - choose.

8. Plot name - does not have a dialogue box, requires only a label.

9. Plot - uses the same dialogue box as Plot area.

10. Information - contains text messages.

11. Make slide - stores the current sketch as a slide. The file will be stored with an extension of .sld instead of .skd. You can store scanned sketches with .sld file extensions. How to create a slide presentation is discussed in the next section of this chapter.

12. View slide - the other half of a slide presentation is to look at .sld files in rapid succession. SKD files must be built and the screen erased manually. This command may be used with Make slide to present sample sketches stored as CD 5.5 through CD 5.7 in rapid succession for a customer, or CD 5.5 through CD 5.11 can be presented to walk the customer through the proposed machine tool design project.

5.16 Slide Presentations

The DLR directory provides a script facility that allows commands to be read from a text file. This feature allows you to execute a predetermined sequence of commands. These commands are contained in a file.scr. A script file for C5.5 through C5.7, might look like:

VSLIDE FIG55	(begin slide show, view C55.dwg)
VSLIDE *FIG56	(preload fig56.dwg)
DELAY 20000	(let customers see C5.5 20 seconds)
VSLIDE	(display slide 2, view file C56.dwg)
VSLIDE *SLIDE 3	(preload C57.dwg)
DELAY 30000	(let customers see C5.6 30 seconds)
VSLIDE	(display slide 3, view C5.7.dwg)
DELAY 10000	(let customers see C5.7 10 seconds)
RSCRIPT	(cycle)

You will recall that these sample were recalled from the LT directory using the SKETCH command and were stored as fig.dwg files. Sometimes, it becomes necessary to show a customer various views of a complex sketch. Suppose that we begin with C5.5 and make several slides of the machine tool sketch. One can be as shown, another can be a PAN right, left, up and down, another can be a series of ZOOM in ,out, and dynamic. This results in 8 slides for C5.5 carried through each sample, that results in 56 slides of the 7 sketches. The 56 slides are put in a script file like shown and you have created a complete slide show by combining the slide feature with the script facility.

In the sketch directory Make slide and View slide submenu items are the same commands as MSLIDE and VSLIDE commands in LT. You can write script files for a sketch directory or another LT directory. A slide is a snapshot of the monitor. You use ordinary commands to obtain the desired view on the monitor and then make a slide of that display.

5.17 Slide Libraries

You can construct slide libraries from .sld files by using the SLIDELIB utility program supplied on the LT support CD. To construct a slide library, enter the command:

Command: SLIDELIB
Library:<fig55/fig56/fig57/fig58a/58b/58c/58d/58e/58f/58g/58 h..>

which places all of the made slides just described into a single file called library.slb. You can now use this and other libraries which contain Shade rendering to produce multiple slide shows called FILMROLL. Now you have a powerful engineer - customer communication format.

5.18 Chapter Summary

Sketching techniques are contained in two directories acad and sketch. The LT directory was used first because we have been using that in the last three chapters. It contained a very limited freehand ability with the SKETCH command and sub command items. It did contain other commands like RENDER, however. This can produce life like sketches. A series of machine tool examples were discussed and were all taken from the sample LT directory.

We then changed directories to find Sketch, a DLR directory product similar to CAD. The Sketch software package was covered so that another sketching technique could be introduced. A third technique called slide making was introduced to show the concept of a sketching library.

6

Lean Image Processing

Lean image processing for AME (Advanced Modeling Extensions) applications is generally thought of as being either passive (use a scanner) or active (use online images) in nature. When an engineer selects an online mode rather than manual documentation that is scanned, a decision is made if the lean models can be displayed from software in real-time. An image that is scanned is referred to as computer-assisted rather than computer-aided. The two terms aided and assisted do not carry the same meaning. Computer-aided is an active process because it appears that the graphics workstation has the ability to solve image problems without prior input by humans. Types of hardware are referred to as dumb or smart. A scanner is dumb, it copies whatever it sees manually drawn or sketched and places it in a drawing file.dxf. A smart device stores drawing files as .dwg, .skd, .sld,.slb, or .scr.

Computer-assisted, passive, dumb hardware devices may be used as much as computer-aided, active, smart devices in an AME environment. If model project files are being developed during the early, preliminary stages of a design, dumb devices are used more. In the final stages of project documentation smart devices are used more. In the last few years real-time graphics and design software packages for engineers have been used. This book makes use of the DLR package called AME which is the CAD application software for mechanical, industrial and electrical engineers. The AME package is written especially for active graphics applications falling into one of four major types: 1) wireform, frame or mesh; see Chapter 10, 2) 3-D surface; see Chapter 11, 3) solids; see chapter 12, or 4) section and profile, see Chapter 13.

6.1 AME Image Processing

The standard products of AME are built into the right side screen menu. You access these menus by clicking MODEL as shown in this chapter heading. You must load AME by the command:

Command: xload ame
Command: DDSOLPRM

This chapter shows how to load the AME for lean modeling and how to use lean image processing (LIP). AME/LIP is done by using the side menu and the pull-down menus MODEL and RENDER. We will study the uses of the remainder menus in the following chapters:

1) Storing Lean Models ------ FILE MENU, 2) Lean Model Geometry --- CONSTRUCT MENU, 3) Vector Model Geometry --------- DRAW MENU, 4) Wireform Generation ----- MODEL MENU, 5) 3-D models –DRAW/3D SURFACE SUBMENU, 6) Solid Model Design-- MODEL & VIEW MENUS and 7) Section Profile Models --- MODEL & MODIFY

The AME software package generates more accurate lean models at a lower cost because LT costs are less than traditional lean modeling packages for large mainframe computers. The AME package handles the repetitive aspects of building models while the user focuses on the creative side of the work.

The problem then is to decide which of the many packages to use at any given time. It might be confusing to give more than one method for model displays, but the truth of the matter is that many packages exist. The same company offers a variety to choose from. DLR, for example, offers extensions for things like COGO (Coordinate Geometry) and

DTM (digital terrain modeling) for Surveyors and Civil Engineers. These are both beyond the scope of this book. The point being that the software chosen for this book is considered the bare minimum for learning lean modeling.

6.2 Simple Image Processing

Before plunging into a detailed description of the many types of AME menu commands that can be used, it seems fitting to devote some explanation to how the popular software choices are used. **SKETCH** is for the architect or engineer who knows very little about computer software or how SCRIPT files are written. This kind of user wants to be able to draw or sketch standard models, saving these as parts for reuse later, as described in Chapter 5 . Therefore, a user is concerned with learning how to use blocks and insert commands.

6.3 LT Lean Image Processing

This second package (without AME) is for the model designer who wants to go farther than the draftsperson and link the image processing with script files which are a form of programming for the output sequencing. We learned how to do this for a slide show in the last chapter. This combination best uses the active part of CAD software. The use of this package requires more knowledge of image processing than either the casual user or a computer programmer. It is true that AME type packages were unheard of several years ago, but today they are common place.

6.4 AME 3D Image Processing

It is much easier for lay people and customers to understand three dimension image presentations. Therefore, I provide this software as a conceptual design tool that let's you create 3D models with ease. It can be used to create a realistic lean model that can be turned so that you can view it from any angle. With this package you can design in 3D and use the software's navigation system to locate yourself anywhere in your model and in space around the model. This is used with the CAD package and data can be passed between the two. The most often used menus are the SIDE and MODEL.

6.5 Using the AME Package

How, then, can an AME package be put to immediate use in the engineering office to do lean modeling? First, pick a package at your skill level. Some companies even provide a zero skill level package called the generic starter kit. It's a CADD course in a box, workbook, CD, and sample models. Just above the starter kit is **SKETCH**, then the genericCAD packages, topped by AME. The purchase costs follow that same stepping.

Second, consider how much time of your daily routine is taken up with simple model shapes or highly repetitive non-thinking tasks. If this is more than one-third, these types of tasks can be done by the LT and release over 30% of the work force to do other design tasks. Thirty percent will return the cost of AME in less than 1 year in an office.

Third, begin by automating the simplest most often used tasks first to gain understanding and confidence in AME usage. If possible, choose a reoccurring model that must be displayed by a different project file each time and requires 1 or more hours to complete by a draftsperson. Chances are the time required to put this task on the LT will also take an hour, but now the task can be used over and over again by inserting different data and displaying it in milliseconds after that. Once a task has been set up on the LT and stored in the controller, the time required to repeat it is 1/100,000 of manual. Naturally one-time jobs are not done on AME software.

6.6 Extruded Models

The simplest model known is produced by taking a 2-D image such as a square and assigning a length along a display axis with the command ELEV. ELEV lets you set the current elevation and extrusion thickness for subsequent entities. A point becomes a line, a line becomes a plane, a plane becomes a solid and a solid becomes the extruded lean model. The first menu item in the pull-down MODEL menu should be practiced with any 2-D combination of points, lines, planes and simple solids until you become comfortable with this first menu item EXTRUDE.

6.7 Revolving the Model

After the extruded model is produced it may be revolved on the display by selecting the REVOLVE or second menu item. This is the easy method to look at your newly created model. You will note that the model appears transparent, this is common of all extruded models. You may select the HIDE command after revolving the model. This will delete all lines behind the visible faces of the model. The model is still transparent it just aids the viewing.

6.8 Using the SOLIDIFY Menu Item

To view the extruded model as a solid use the third menu item within the pull-down model menu. In order to illustrate the same function, a point, line, plane and solid are again shown solidified by menu selection. A point can not be solidified but it can be changed from a 2-D location to a location within a 3-D lean model and that is what is done quite often. Likewise a line can be changed from a 2-D entity into a model edge, while a plane becomes a model face.

6.9 Working With Non-extruded Models

When you select one of the primitives from the fourth menu item, it may be an extruded model primitive such as BOX or WEDGE. Most likely it will be one of the other non-extruded primitives. These primitives maybe selected from this menu item or by called them directly by their command names. These are: BOX, PYRAMID, DOME, SPHERE, WEDGE, CONE, TORUS, REVSURF, RULSURF, EDGE, TABSURF, SETSURF and MESH. As you begin to construct various lean models you will find other primitives that you can create and store under BLOCK names to be used as easley as those just described. Many primitive shapes are used in final models. These are put together using the next section of the MODEL menu. You may join two or more primitives with the UNION menu item. You may want to take apart joined primitives with the SUBTRACT item. You may separate non-joined primitives or other sections of lean models as you see fit. In some cases you may even want to intersect two sections of a model.

But in most cases you will want to do a number of model modifications. You will notice that in the MODIFY > menu item within the MODEL pull-down menu has a sub-menu. Inside this you can do things like move sections of models.

6.10 Model Modifications

The Modify > sub-menu provides for moving model parts, changing the shape and size of primitives. You may separate without subtracting and create a new primitive. A useful menu item is called CUT SOLIDS, we will need this ability to create section and profile models for Chapter 13. Chamfering of solids is used many times in creating lean models. Suppose that we needed to make a pulley model? We could begin with a chamfered cylinder, reduce its thickness with ELEV, place a straight sided cylinder directly on top of it with a UNION command and then place another inverted chamfered cylinder directly on top to finish the pulley lean model. Many applications also exist for rounded and filleted edges on lean models. Use the FILLET SOLIDS menu items for models requiring rounded edges.

6.11 Model Setup Menu Items

Basic setup information may be given before, during or after a model is completed by selecting the engineering units required, such as English, SI or CGS. These are selected from the variables setup submenu. The upgrade variables will allow you to work with different versions of CAD and AME. If you already own a CAD package then all you will need to do is purchase the AME extension package. Double Precision changes from 8 place accuracy to 16 place calculation mode, use the command SOLUPGRADE. The last thing you will need to do is set the AME script compatible with your present version of CAD, use the command SOLAMECOMP.

6.12 Model Inquiry Items

The Inquiry > menu item is used to list properties of solids/regions by using the SOLLIST command. To list the mass properties of lean models use DDSOLMASSP. You may calculate areas for solids or regions by using SOLAREA. Calculate the interference between two nested models by using SOLINTERF. You can set the variable that controls mass decomposition direction with SOLDECOMP. And finally you clan set the variable that controls mass subdivision by SOLSUBDIV. A region is often used instead of the entire model, for example, the cut surface of a model can be studied separate from the body of the model.

6.13 Model Display Techniques

The submenu items known as MESH can be used to display models.. Before a surface can be shaded, its topology or roughness must be known. The best way to show this is place a mesh or grid over the entire surface with SOLMESH. If the entire model is to be shaded then the Wireframe menu item is selected by SOLWIRE. You may set the wire density with the command SOLWDENS. You may copy a model feature such as a boss or counterbore with SOLFEAT. This is helpful with repetitive holes, slots or other part features. You can even copy a part profile with SOLPROF.

6.14 Utility Commands

The submenu items called Purge Objects is a utility command. When a lean model becomes highly complex as you will need to remove or erase all unnecessary entities such as meshes, wireframes and so forth. This should be done after rendering however. In addition to the model purge, you select from the following:

```
Associate a material for the model ---------DDSOLMAT.
Set the UCS on a model face --------------SOLUCS.
Import a Solid file ----------------------------SOLIN.
Create a Solid file ----------------------------SOLOUT
Load AME -----------------------------------LOAD AME.
```

6.15 Rendering the Model

The pull-down menu RENDER gives you a choice for rendering. Rendering means creating a realistic looking lean model image of a wireframe using all the geometry, view, and light information for that model. The RENDER command will render the current wireframe model. By default (if you have not selected a scene or view) the command uses a preset value for view, light and scene. To set these yourself see the RENDER Preference dialogue box.You may render any wireform (mesh surface) or wireframe (3-D model) with this command.

6.16 How to Shade a Model Surface

The shading technique used for SHADE commands in AME is easy to use. This command lets you produce a shaded lean model without consideration for view, light and scene. It will produce an image that removes hidden lines and displays the faces in their original color with no lighting effect. After the SHADE command is used, the screen goes blank for a period of time, the length depends on the model number of faces. You may speed up this process by using HIDE first.

6.17 How to Hide Model Faces

The HIDE command is used to take away faces on the model that are not visible to the viewer. You may use this command to speed up others such as render, shade, vpoint or dview. Whenever a wire model is produced on the screen by selecting any of the model parts or techniques, all lines are present. This includes those that would be hidden behind other faces. To eliminate these hidden lines just type in the HIDE command. This is an important command because we often want to rotate the model to study all sides. The model will appear to be solid if this command is used throughout the image display process.

6.18 Views

This menu is so complex, with so many variables to be set, that it is udsually done by dialogue box. You may fill the dialogue box by using the DDVIEW command. Fill the dialogue box with care. Selecting too many entities, slows dragging and updating of the image. On the other hand, choosing too few may not provide an adequate view. The recommended ones are:

CAmera/TArget/Distance/POints/PAn/Zoom/TWist/CLip/ Hide/Off/Undo/X

CAmera rotates the point from which you are looking around the target point. Move the CA with the graphics cursor in the graphics area of the graphics tablet. TArget rotates the target point about the CA. The effect is like turning your head to see different views of the model from new vantage points. Distance moves the CA in or out along the line of sight. This turns on perspective viewing. POints lets you locate the CA and TA using (X,Y,Z) model coordinates. You can enter these by object snap or X/Y/Z point filters (see Chapter 14).

PAn shifts the image without changing the zoom from side to side or up and down. Zoom has two modes, perspective and normal. TWist lets you twist or tilt the view around the line of sight. CLip allows front and back clipping planes (invisible walls) that you can position, perpendicular to the line of sight. All lines behind the back are clipped or removed as are all lines in front of the clipping plane. Hide performs hidden line suppression. Off disables front and back clipping. Undo reverses the effects of the last operation. eXit ends the viewing operation.

6.19 Lights, Scenes and Finishes

Light inside the RENDER menu item is used to include light location and light targets. You use the LIGHT command to add a new light and modify or delete an existing light source. The lights dialogue box lists all lights in the current lean model image display. The scene for a lean model image display is set by a dialogue box. The scene can be modified. Finishes can now be selected.

6.20 Selecting Rendering Preferences

Use the command PREP to display the dialogue box. From this you can select either full render options or quick render options.

6.21 File Storage

Storing lean images as files is covered in detail in Chapter 7 of this book, in this chapter we will need to know how to use the Statistics and Files menu items. You begin by selecting the REPLAY dialogue box. After you select an image file in the REPLAY box above and pick OK, the image specifications appear in a dialogue box.

6.22 Unloading and Ending Renderman

Renderman is the software collection of LISP routines that carry out most of the functions of the RENDER and MODEL pull-down menus. It is possible to speed up the processing times of early lean building and construction techniques that do not require shading or rendering by ending or unloading the rendering routines. This should be done at the end of a modeling session after the completed lean model is in its published form.

6.23 Chapter Summary

The purpose of this chapter has been to introduce AME (advanced Modeling Extensions) package that changes the appearance of the LT screen so that lean models can be processed in a very simple and direct way.

You were shown how to load AME, how to produce simple lean objects or parts that could be used in more complex models and how to process these images in a meaningful way. The AME package is written especially for active graphics applications falling into one of four major types: 1) wireform, frame or mesh; see Chapter 10, 2) 3-D surface; see Chapter 11, 3) solids; see chapter 12, or 4) section and profile, see Chapter 13.

7

Storing Lean Models

The basic technique for storing a finished lean model is essentially the same as those computer generated files we have created in the first six chapters of this text. Whenever we save something, a file is created. Soon we have hundreds of files running around each containing special symbols, conventions, lean constructions and manufacturing details. The purpose of storing models is related to the concept of layering.

Different layers are the contributions of a team of specialists, which include the engineer (data preparation), designer (product design), detailer (image processing) and illustrator (renderman). In some cases one person may often function in several capacities and develop ideas for more than one layer, particularly in a small office. On large projects, the practice is to isolate design functions and activities to a single layer.

7.1 The Storage Technique

Let's examine Table 7.1 to study how the computer storage works for a completed lean model. The finished rendered model is stored in a separate locked directory. This way nothing can be added, deleted or altered without first entering that directory. Think of the directory as a central hub, which can access preliminary sketches, working production drawings, dimensions and specifications, all the manufacturing details including production schedules and planning sheets. This is the usual database relationship for the contents of an engineering project.

Think of the storage concept like this, there is a single piece of hardware called the controller (public library). Inside the controller are several directories (shelves in the public library). Inside each directory are several files (books on the shelf). Inside each file there are layers (pages in the book). Each layer may have one or more viewports (written paragraphs and pictures on the pages). We store related lean model details in separate viewports on the same layer. We store similar details on separate layers in the same file. It sounds confusing, but it is easy to use as:

Lean Model Storage

7.2 Examples of Lean Model Directory Files

An example of separate stored layers within a single file are as follows:

1. Data description for the lean model
   A. XYZ filters,
   B. Descriptive geometry,
   C. Vector geometry,
   D. Design analysis.
2. Wireform presentation
   A. Working views with dimensions,
   B. Model display with dimensions,
   C. Tolerances and notations.
3. Solidification of the model
   A. Hide and
   B. Shade
4. Rendering the finished model
   A. Lights,
   B. Scenes, and
   C. Finishes.

Each of these layers is stored using a viewport approach. If you have multiviews of the same object, for example, put the plan view in one viewport and the elevation in another viewport. They can then be stored on the same layer if desired.

7.3 Use of layer viewports

Because of the extensive education and experience with viewports and layers, the engineer and model designer act as team leaders with the advice and representation of the product customer/owner throughout the storage process. This process was discussed in detail in Chapter 2. In this case the robotic lean model's mechanical group leader is responsible for the planning and storage of the robots parts, hydraulics and other mechanical systems. The electrical group leader produces the storage for the necessary circuits and ports for power, lights, fire alarm and clock systems, and the communication networks.

The team leaders, under the guidance of the engineer and approval of the owner, prepare the viewports to begin the layers as needed. The layers are stored in common files as needed. From these data, detailers will prepare drawings and diagrams for the entire project. The concept of viewports on layers within files and their speed of production are extremely important since this system, if properly executed, will reduce the costs for the entire robot design program.

7.4 How to Control Multiple Viewports

The viewports command VPORT lets you control the number of viewports on a layer. It also provides facilities for saving and restoring named viewports. The command format is:

Command: VPORT
Save/Restore/Delete/Join/Single/?/2/<3>/4:

The menu bar choices are:

S - lets you assign a name to the current viewport (the number and placement of active viewports and their settings) and save it for future use. You can save any number of named viewports with the layer, to be recalled at any time. R - restores a previously saved viewport. D - deletes a named viewport. J - merges two adjacent viewports into one rectangular viewport. SI - use this to return from a joined viewport. ? - displays the ID number and screen location of the active vport. 2,3,4 - divides the current layer into two, three or four viewports.

7.5 Dragging Details Between Viewports

When multiple viewports are active you may highlight (capture) a detail and move it into another viewport. You highlight a detail by grouping all its objects (graphics and text) into a single display object. This can then be moved between viewports and around the layer. This is helpful and is like a cut and paste within a word processor.

7.6 Use of Multiple Layers to Create a File

You can use the layer command to create new layers, select the current layer, set the color of a layer, indicate the linetype, turn layers on or off, and list the defined layers within the file. The command is used as:

Command: LAYER
?/Make/Set/New/ON/OFF/Color/Ltype/Freeze/Thaw/LOck/Un

The menu bar choices produce:
?

Layer name	State	Color	Linetype
0	On	7 (white)	Continuous
Titlestrip	Off	7 (white)	Continuous
FL-1-plumbing	On	3 (green)	Dashed
FL-2-walls	Frozen	12	Continuous
FL-3-plumbing	On	5 (blue)	Continuous
FL-4-walls	On	1 (red)	Dot
Foundation	On	240	Continuous
Dimensions	Off	7 (white)	Continuous

M - establish a layer and make it current.

S - select current layer.

N - create a new layer.

ON - turns layer on.

OFF - turns layer off.

C - 1=red, 2=yellow, 3=green, 4=cyan, 5=blue, 6=magenta, 7=white.

L - continuous, dashed, center, dot, or dashdot.

F - entities are not displayed or plotted but data is stored on layer.

T - undo freeze.

LO- keeps anyone from adding or deleting information

U - allows anyone to add or delete information

7.7 Directory Development and Storage Stages

Most lean modeling projects have four storage stages:1) Preliminary planning and sketching, 2) Storage of viewport details, 3) Preparation of specifications, dimensions or descriptions, and 4) Scheduling and management of the actual prototype or testing model construction.

Further project advantages exist with layering because the engineer or model designer will always have details other than those of geometry construction, for example, plans for the use of vector analysis, surface development such as smooth faces and mesh analysis, storage of part feature materials, and the like. To assist the designers in storing these details let's go through the four stages.

7.8 Preliminary Planning and Sketching

Refer to Table 7.1 again. Notice that the preliminary sketch files have an arrow indicating information flow into the project directory. The project directory is empty before the preliminary studies begin. Read Chapter 5.2 through 7, preliminary sketches. Notice that the preliminary sketches can produce presentation rendered models for communication with the customer/client. If these are not successful, then a project directory will not be needed as the product design project was never approved for purchase.

7.9 Storage of Viewport Details

The preparation of viewport details for directory storage in a discrete layer involves two steps.

1. A rough sketch is prepared by the engineer or model designer and saved in file.skd, the procedure for this was described in detail in Chapter 5 of this text.
2. The person responsible for the detailed model description refines this skd file into a finished display that is sent to checking, approved and stored in the directory as file.dwg.

7.10 Preparation of Specifications and Dimensions

You will note from Table 7.1 that information flows in both directions, into and out of the project directory. The dimensions block is shown before the specs block because sizes, shapes and limits are needed before a set of specifications can be developed. For example, in a set of specifications the General Conditions for a set of computer generated specs are:

The latest edition of the standard form of general conditions of the contract published by the Society of Manufacturing Engineers for the design of robotic elements shall be understood to be part of this specification and shall be adhered to by this model designer.

The limits for model projects are spelled out in the Special Conditions section of the specs as:

Sec. 1. EXAMINATION OF CUSTOMER'S INDUSTRIAL SITE. It is understood that the designer has examined the owner's site and is familiar with all conditions which might affect the execution of this contract and has made provisions therefore in the bid.
Sec. 2. TIME FOR COMPLETION. The MODEL shall be completed within 100 calendar days after written notice to proceed is insured.
Sec. 3. LIQUIDATION DAMAGES. The designers and the sureties shall be liable for and shall pay to the owner the sum of five hundred

($500.00) dollars for each calendar day of delay until the model is completed.

Sec. 4. EXISTING Tools. Existing tools within the owners.......

Sec. 5. TESTING. The engineer in charge shall

Sec. 6. PROTOTYPE GUARANTEE. The acceptance of this

The general scope of work for the computer generated set of specs includes the following divisions:

1) Model design sketches and planning sheets, 2) Preliminary rendered lean model, 3) Wireform, fully animated slide presentation, 4) Regional section views, articulated slide presentation, 5) List of newly created primitives in model geometry, 6) Slide presentation of vector analysis robotic elements, 7) Schedules of prototype testing and results, 8) Manufacturing notes and working drawings, 9) Finish hardware, 100 Hydraulics, plumbing and electrical, 11) CADAM reports, and 12) Promotional colored renderings of the robot.

7.11 Scheduling and Project Management

Look at Table 7.1 again, notice that this group of activities are all ringed to the left hand side of the diagram. Information flows from the directory and between the manufacturing details and the actual schedules. The Directory holds the CAD SQL Extension (ASE) menus for schedules of:

1. Electrical and mechanical utility diagrams can now be created and stored in the directory. Sections of the CADAM report are added to the model directory.
2. Wireform models of robot parts are filed so that a XYZ filtered dimensioning system is stored in the directory. This information can be used to create manufacturing diagrams and schedules.
3. Solid models of a part are stored here. Vector analysis can be completed and stored in the directory.
4. Rendered model parts are stored here. These types of presentations are used in the promotional printed materials.

7.12 Storage of the Model Displays

We are now ready to complete the storage process shown in Table 7.1. It is the right hand side of the diagram which contains the lean model modeling process wireform through solids.

1. Wireform - is used to assist in the constructive geometry and design of all 3-D entities. It is a series of stored layers consisting of a 3-D view of the design model. This shows the model sizes, various distances from reference and boundary points, locations of plant columns, and other industrial site details. See Chapter 2.
2. Regions or sections - usually taken vertical of the whole model to show the relationship between operational systems.
3. Primitive elevations - are either internal or external views of the robot projected on a vertical plane. Normally four layers or four separate files are used to store this information.
4. Solids - are discussed in Chapter 12.

We are now ready to store an entire directory once the dimensioning is complete (see Chapters 14 and 15). These files must provide enough information so that along with the written specifications, schedules and legal contracts no design decisions are left to the manufacturer. The directory now contains all the necessary items shown in Table 7.1.

7.13 Chapter Summary

Stored lean models are extremely useful in the industrial market place and therefore are subject to the latest technology advancements with regard to how they can be used. Suppose a client can access information about a project over the internet? Would this have an affect on project approval? Suppose you provided the client with a slide show burned onto his own CD's? Would this have an affect? Why mail design drawings when you can transmit information and images electronically?

These questions should start the what if we did this with our clients. This chapter is needed at this point in this book to show how powerful lean data can be, how to protect it, and how to manage it. Storage techniques were discussed from the standpoint of single project directories, multiple files made up of multiple layers, created out of multiple viewports. This useful concept can save numerous work hours within an engineering design office.

8

Lean Model Geometry

Lean model geometry is contained in the AME pull-down menu, CONSTRUCT. In this chapter you'll see how 3D commands like; array, face, mesh, and poly are used to copy and mirror lean images. During construction you will often need to use the divide, measure, and offset menu commands. When the geometry construction is complete you can place it in a block. Before constructive geometry can be useful, we must understand what it is, what it is capable of, and the purpose it serves within this reference book. It is a mixture of geometry (mathematics), 3D computer-aided engineering graphics and digital computing. It is not orthographic projection based upon descriptive geometry which is a mapping of 3D onto a 2D surface, usually a piece of paper or monitor screen. Lean model geometry, as we use it today, evolved from the transition into the electronic age.

8.1 Selecting the Display Options

In order to select the proper display options we must understand space coordination. These are known as the Euclidian Postulates for space.

EP1 There is a set (group) of points (real numbers, mapped as ordered triplets). Certain subsets (members of the group) are called lines; at least two points are needed to define a line. Certain other subsets are called planes; at least three points are needed to define a plane, and solids contain at least four points not in any one plane.

EP2 There is exactly one plane containing any three non colinear points.

EP3 Each plane satisfies all other postulates.

EP4 Each plane separates space into half-spaces with the following properties: (1) if two points are in the same half-space set, a line between them does not intersect the plane, and (2) if two points are in different half-spaces, a line between them will intersect the plane.

From these few postulates all the theorems of modern lean model geometry can be derived and provided as display options. However, we will not derive any theorems in this book! All that is needed is an understanding of the elementary aspects of space geometry and the ability to visualize and display 3D CD figures. We will see later in this chapter further similarities: (1) parametric descriptions of a line in space are quite similar to an equation of a line in a plane, (2) distance descriptions are simple extension of the equation for the plane, (3) planes in space (faces) can be described by the location of three or more lines, and (4) solids can be described simply by means of equations.

With this background then the display options to be studied are POINT, VPOINT, From Point, Insert Point, PLINE, POLYLINES and 3D (FACE, MESH, POLY, EDGESURF, REVSURF, RULESURF and TABSURF).

8.2 Point Options

We have defined Euclidian space and have given it coordinates. But often in the application of mathematics to problems of engineering and computer science it is the converse situation that prevails. We are given ordered triplets of real numbers and use space to illustrate graphically the relations between the ordered triplets. In some cases these is no need to choose the display option; often the display software has only one method for graphic display. Here a point (P2) is:**P2 (1,3,2).** Furthermore, there is no definite way to orient the display axes, so we use the display option called VPOINT. For ease of observation you may reposition the axis tripod by clicking on the compass icon and moving the target + within the compass icon. You will note that the display axis tripod is dynamic on the display monitor allowing you to position the geometry output into a convenient viewing position. Many of the lean geometry menu items that appear below the command line on the CAD display have sub-commands like from point or insert point as:

Command: Block
Block name:
Insertion point:

The distance between points can be determined by using the CAL command. we are provided P1 = (X1,Y1,Z1) and P2=(X2,Y2,Z2). We choose to add point 3 so that points 1 and 3 have the same Z value. The distance P1P3 is given by the equation in the plane, namely P1P3 = ((X1-X2)**2 + (Y1-Y2)**2)**.5.The points P2 and P3 are on a straight line and P2P3 = ((Z1-Z2)**2)**.5 This makes P1P2P3 a right triangle, and by pythagorean theorem, P1P2 = ((X1-X2)**2 + (Y1-Y2)**2)**2 +(Z1-Z2)**2)**.5

8.3 Line Options

There are several conventions that are observed in lean model geometry dealing with lines (parallel, intersecting and perpendicular). Coplanar lines that cannot intersect however far extruded are called parallel lines. Through a point, only one line can be displayed parallel to another. Two coplanar lines are parallel if both are perpendicular to a third coplanar line. Two parallel lines to the same line are parallel to each other. Finally, two lines perpendicular to the same plane are parallel to each other.

If two coplanar lines are not parallel, they will intersect if extruded and two straight lines can intersect at only one point, called a concurrent point for intersecting lines. Perpendicular lines are lines that intersect at right angles. An infinite number of 3D lines can be displayed perpendicular to a given line. However, from a given point only one line can be displayed perpendicular. The most common way to display a line is:

Command: PLINE
From Point:
Arc/Close/Halfwidth/Length/Undo/Width/<Endpoint of line>:

We notice that a PLINE through three noncolinear points will produce path C1. This is displayed with the PLINE/Arc option. If lines are used to define the edges of planes, use the PLINE/Close option. The lean geometry software used for this chapter lets any 3D line have starting display coordinates of X0, Y0, Z0 and ending coordinates of X1, Y1, Z1. Next the line distance (D) is determined. In order to do this we need the direction angles as (1) the X-direction cosine, (2) the Y-direction cosine and (3) the Z-direction cosine. If we label these angles C1, C2 and C3, then the following equations can be read by the display software: C1 = (X1-X0)/D and C2 = (Y1-Y0)/D and C3 = (Z1-Z0)/D if C1, C2 and C3 are the direction cosines of a line, then: C1**2 + C2**2 + C3**2 = 1, given the parametric description of cosines C1, C2 and C3, let X1, Y1 and Z1 be a point on the line at distance D then: X = X1 + C1*D and Y = Y1 + C2*D and Z = Z1 + C3*D.

8.4 Plane Options

Various types of plane surfaces (point/plane, line/plane, plane/plane and parallel) are used in lean geometry. A plane defined like these can be displayed by any two lines within the plane surface. The lean geometry completely describes the projections of these lines on the coordinate plane surfaces. All lines lying in a flat plane surface are called coplanar lines. All 3DFACE generated planes contain coplanar lines. Not all surfaces are planes, but all faces are planes. A surface may be either a plane (face) or sculptured surface.

There are four main display options for surfaces on lean models that are constructed from AME software. They are: EDGESURF used to define the edges of holes, slots or other part features, RULESURF defines a surface with clean smooth line segments, REVSURF is perhaps the most useful in industrial applications. And finally, TABSURF which is used to define the surfaces of cylinder pins, and most extruded models.

8.5 Solid Options

In Table 8.2 the display of solid options from lean geometry (LG) are listed. The LG solids are divided into wireform and lean. Wireforms are covered in Chapter 10, while the remaining LG leans are divided into sphere, ellipsoid, paraboloid, hyperboloid, and torus. Sphere and torus are provided as primitives as seen in Chapter 6. Ellipsoids can be displayed as two axis equal, general or three axis unequal. Paraboloids can be displayed as revsurfs or elliptical. Hyperboloids are displayed as revsurfs or ellipticals.

There are four common type solids used in lean geometry displays; 1) a solid composed of faces, 2) a combination of faces and surfaces, 3) all sides are surfaces and 4) a single surface is used.

8.6 Menu Options

The menu options for LG are ARRAY, 3DARRAY, COPY, 3DMIRROR and BLOCK. The construct menu options for modifying LG are Mirror, Chamfer, Fillet, Divide, Measure, and offset. Let's examine each of these menu options and study example uses for lean geometry.

8.7 Using the Array Options for Lean Models

The polar option of 3DARRAY was used to in the completion of the lean model stored under rotor.dwg. This model was created by joining with the union command three primitives Box, Pyramid and Cylinder. Next copies were made with the COPY construct menu item and these were displayed with the 3DARRAY display option.

8.8 Using the Copy Options

Copies of the joined primitives; box, pyramid and cylinder were copied in order to be displayed in a polar array. The polar array was stored as a solid by the use of the HIDE command before it appears as a composite model rotor.dwg. Many examples exist for the use of copy, we use the lean geometry technique to create a helix path and copy this path three times to produce a new composite model drill.dwg. Copy can be used with the SURF commands. For example, ruled surfaces can be copied to produce other tool images such as the reamer.

8.9 Using the Mirror options

The composite model primitives are mirrored so that the left side of the rotor.dwg could be displayed. We used a combination of copy and mirror3d to display a screw thread profile. If the combinations of options can then be used to display common threaded industrial items..

8.10 Using the Modification Options

The modification options of divide, measure and offset were shown for the joined composite model primitives. Before the composite was assemblied it needed to be measured to see how many copies were needed to complete the polar array. To do this the primitives were divided into equal parts and scaled by the use of offset.

When the primitives are joined, they may then be stored or inserted as a block. Additional modifications may need to done before storing the composite model. The sharp edges and corners may have to be finished with either a chamfer or fillet. Fillets and chamfers are used along with a helix to model a fluted cutter head. Other tools may be displayed in this manner.

8.11 Chapter Summary

Lean geometry for model construction was introduced as contained in the AME pull-down menu shown in Table 8.1. In this chapter 3D commands like; array, face, mesh, and poly were used to copy and mirror lean images. During construction we used the divide, measure, and offset menu commands. When the geometry construction was completed we placed it in a block.

Constructive geometry was found to be very useful, because we now understand what it is, what it is capable of, and the purpose it will serve within this book in later chapters.

9

Vector Model Geometry

Vector model geometry is the study of graphic statics, model velocity and acceleration analysis. This infers that lean models are structures that withstand stresses. The stresses are either pushing or pulling on each member. A pushing stress is called compression and has a + sign. A pulling stress is called tension and has a - sign. If the stresses are stationary, static vectors are displayed, if the stresses are in motion, dynamic vectors are displayed. In either case, the application of statics or dynamics to the solution of model structures is not new and has been in use by engineers for a number of years. The addition of lean modeling has revolutionized how we use vector model geometry. In the first eight chapters of this reference book, the illustrations that accompany the text, are provided free of charge at the mailing address indicated. The illustrations for the remaining chapters are all laptop viewable video clips. It is not possible to diagram motion inside a book. Books are wonderful collections of knowledge, but they are static not dynamic. It is not possible to place a model under a polariscope and twist it in real time and show the results as a single picture. A CD video clip of what happens is the best way to illustrate vector modeling. I will be describing some of the video clips to you as I view them for this chapter. Not an ideal situation, you will grasp that after you view your first vector model. If you are familiar with polariscopes and the colors they produce, please skip this chapter and go directly to Chapter 10.

Like lean geometry (LG) of the last chapter, vector geometry (VG) contains powerful tools for problem solving. When the AME equipped LT is used to automate this problem solution, we call it computer-aided VG. CAVG representation of the forces that act in various members of a lean model structure posses many advantages over manual solution; the primary advantage, beyond presenting a picture of the stresses, is that most problems can be solved with the speed and accuracy of the computer.

Combined with the skills outlined in Chapter 8, stress may be presented much more accurately than the various members can be sized, since in sizing we must select, from a handbook, members capable of withstanding loads equal to or greater than the design load. Using VG a model designer computes a size and then applies a factor of safety when designing any lean model.

Customarily, problems in statics or dynamics are solved by manual algebraic methods. In this chapter we will assume that you are familiar with one or the other. With no previous background, you will gain little from the study of how to automate it. Before proceeding further; study the screen output shown in the chapter heading which compares LG and VG display methods, notice the items in the pull-down menu. We begin the study of VG by building directly on the skills learned in the last chapter.

9.1 Vector Notation

In Chapter 8, line segments were labeled starting point and ending point because they were not vectors. They were known as scalar values. In this chapter we deal with line segments know as vectors. This is the name applied to a line of scaled length that represents the magnitude and direction of a force. An arrowhead placed on the line, usually at the end, shows the sense (which way the force acts). The next noticeable difference is that vectors use Bow's notation for labeling as shown in the chapter heading as the space between vectors; A, B, C, D, E and F.

You will note that the space between the vector has been labeled so that vector AB is the first vector shown. Bow's notation is not vector notation. Bow's notation is the labeling of a display diagram to read around two or more vectors called joints. The reading is made clockwise around the joint. Vector notation looks like:

$\longrightarrow$
AB = - 45 | 11,22,33

and is read, vector AB is 45 tension vector units long (magnitude) and the angles from the display XYZ axis are 11,22 and 33 radians.

9.2 Vector Addition and Subtraction

C9.1 illustrates model joints. These joints are shown in lean model form and then again as vector space diagrams. We can use these model joints in the space diagram to demonstrate vector addition and subtraction. The parallelogram method of adding vectors ABC and subtracting vectors DEF is shown. The addition of vectors is called the resultant, while the subtraction is called the equilibrant of the joint. They are both the diagonals of the parallelograms displayed the resultant points away from the joint and the subtraction points towards the joint. In other words the resultant (sum) is the total replacement vector for the joint and the subtraction is the equilibrant of the joint.

9.3 Resultants

If a group of vectors (more than two) that describes a system of joints are in equilibrium, the joint is said to be balanced. The resultant of this system is always zero. Systems are either balanced or not balanced. Balanced systems are static (not moving). Unbalanced systems indicate velocity (constant motion) or acceleration and deceleration (changing motion speeds). Vectors represent abstract quantities for lean models. Two or more vectors acting together are required to describe a system. To solve a problem in which vectors are displayed, a resultant is often found. It is found by the use of two diagrams. One is called the space diagram; it shows the relationship in the physical system and indicates how the vectors are applied. The second is called the stress diagram and is built from the space diagram to determine characteristics of the system, for example, balanced or unbalanced.

Suppose we look at a simple two vector diagram shown in C9.2. If the two vectors act simultaneously at point A, the result will be a path shown as dashed from A to C in this case the addition or subtraction can be shown by a dashed line, an arrowhead at the right side indicates resultant (sum) and an arrowhead at the other end indicates equilibrant (subtraction). If the force vectors act separate and not together, the path taken will be A to B to C along the vectors. If, however, the forces act together, then the path is A to C. In C9.2 and 9. 4 and 9.5 the joints acted upon in a concurrent manner, in other words they acted from a common point. In C9.1, the forces acted in a non-concurrent manner. In this case a resultant and an application point will found as shown in Section 9.6 of this chapter.

9.4 Resultant of a Concurrent System

When more that two vectors make up a system of forces and they are concurrent as in CD 4 or the individual vectors act from a common point making it a concurrent system, they also lie in the same plane making it a concurrent coplanar system of vectors. C9. 4 also represents a concurrent coplanar forces of given magnitude and directions. The system of vectors is displayed in a space diagram and a stress diagram. The stress diagram represents the vectors in the space diagram connected end to end at the same angles of application. We began with the first vector F1, from the space diagram and connected F2 to it at the determined angle of 60 degrees. Each vector is then connected, tip to tail, and this fashion we have all five vectors displayed as the stress diagram.

After the stress diagram has been displayed we may want to find the single force that will have the same effect as the five vectors, it is displayed as R = 31 @ 28* line in the example. This line may be either a resultant or equilibrant depending upon the placement of the arrowhead. If a stress diagram closes, it is balanced and has no resultant.

If however a balanced system is desired then reverse the arrowhead along the line.

9.5 Resultant of a Noncoplanar System

Noncoplanar systems are as easy as coplanar systems with constructive geometry skills. Remember the vector notation in CD 9.3 gave us the angle of rotation around each of XYZ display axis. The space diagram shown in C9. 4 would look like this for a Non-coplanar system and the stress diagram would consist of four parallelograms.

9.6 Determining Resultant Point

The general term for the process of replacing a group of vectors by a single vector is combination or composition. This process is the addition of two or more vectors. The opposite process, that of replacing a single vector by two or more vectors having the same effect, is called resolution Each vector in the new system is called a component of the given vector. Resolution and composition are useful techniques when studying noncon-current systems of vectors.

In C9.5 we have a noncurrent system of vectors. By using lean vector geometry we can connect (add) these vectors tip to tail. This clearly shows that a resultant is present. The magnitude of the resultant can be computed and displayed, but the point of application has not been found. To do this locate a convenient point beside the example shown. This point is called a pole point. From the pole point, display construction lines to each tip and tail. We now have construction lines; OA, OB, OC, OD, OE and OF. They will be used to determine the point of application. Using constructive geometry, place these lines at the same space angles into the space diagram shown.

9.7 Equilibrants

If a single vector is added to an unbalanced system to produce equilibrium, that vector is known as an equilibrant. The equilibrant in an unbalanced system will always have the same location and magnitude as the resultant of that system but will have the opposite sense. Many statics problems can be worked because an equilibrant can be added to a system or the system is already in a state of equilibrium.

In C9.11, let us assume that the wheel in the space diagram is to pushed over the 6-inch block: The horizontal force tending to push the wheel is applied level with the centerline of the wheel. A stress diagram of the concurrent system can be displayed because we know the system is in equilibrium. The stress diagram indicates the direction and magnitude of the forces in the balanced system. As shown in the space diagram, the lines of action of the three forces meet in a common point, and the stress diagram closes. A slight increase in horizontal force will produce motion.

The next question that might be asked is whether or not the center line is the most ideal place to apply a pushing force if labor saving is important. Two other points of application are selected.

9.8 Velocity Analysis

The term velocity is used synonymously with speed in lean geometry. This is incorrect in vector geometry, because velocity includes direction and sense as well as speed. The linear velocity diagramed in CD 6 is not fully identified until the direction and sense in which it is moving and the rate at which it is moving are found in the VG diagram. Therefore, velocity is shown in C 9.6 as either uniform speed plus direction and sense or variable speed plus direction and speed.

9.9 Velocity Display Scales

In the VG solution of problems related to velocity, it is necessary to display the lean model element full size, smaller or larger scale. This display scale is expressed in three ways:
1. proportionate size (element size is multiplied by a factor,
2. the number of display positions on the monitor equal to a metric unit on the lean model element, and
3. a metric unit equals so many units on the model element.
The display scale is designated S(k). The velocity scale, designated S(kv), is defined as linear velocity in distance units per unit of time represented by 1 display unit on the monitor. If the

linear velocity of a point is 5mm/sec and the S(kv) is 5, a line 1 unit long on the monitor would represent a linear velocity of 5mm/ sec and would be written as:

S(kv) = 5

9.10 Resolution and Composition

As described in section 9.6, the process of obtaining the resultant of any number of vectors is called vector composition. The reverse process of breaking up a single vector into two components is vector resolution.

A vector resolution is shown video clip C 9.1, by displaying two components parallel to scalar lines making any desired angle with each other. In other words, the sum or original vector will be the diagonal of a parallelogram obtained with the resolved vectors forming two of the sides The resolution process, shown begins by displaying a resultant vector R. R is composed of vectors A and B. A and B can be resolved so that a system can be displayed.

Obviously, this example is for illustration only, because the model designer selected parallelograms at will. Resolution is a valid technique for the analysis of velocity polygons as shown in C 9. 6. For example, if the velocity of one point and the direction of the velocity of any other point on a lean model element are known, the velocity of any other point on the element may be obtained by resolving the known velocity vector into components along and perpendicular to the scalar line joining these points.

In another example we begin by displaying the two points on the lean model member, shown as A and B. The velocity of point A can be displayed as a vector, V(A) in the example. The direction of the velocity of point B is known and can be displayed as a scalar line, called L in this example. Since this is a lean model, the distance between A and B is constant and a scalar line can be displayed through them as shown. Use the construction scalar to resolve V(A) into components along and perpendicular, as shown in this example. Use the vector labeled ALONG to extend the construction scalar display of AB and drop a perpendicular vector display at the tip of ALONG to meet the direction line L. This is the perpendicular component of the velocity of point B.

The meeting of this component with the direction line L is the total velocity of point B, labeled V(B).

In a more useful sense, suppose that a lean model member has three points, labeled A, B and C. The velocity of A is known, the direction of B is known, but neither the magnitude nor the sense of pint C is known. Using the techniques of resolution and composition, the total velocity of point C can be found. Begin the solution process by displaying the physical location of the three points, the velocity of point A and the direction of point B. Display the constant scalars through A, B and C shown. Next resolve V(A) into components along and perpendicular to scalar AB as in C 9.23). Display the along component at point B and drop a scalar connector to line L as indicated. Resolve V(A) into components along and perpendicular to scalar AC, as shown. Display the along component at point C and drop a scalar connector called M in (24). Resolve V(B) into components along and perpendicular to scalar CB. Display the along component at point C and drop a scalar connector called N. This is the total velocity of point C, labeled VC).

9.11 Velocity Display Axes

In CD 6 the lean model has elements rotating about a fixed axis and a moving axis. To Display this, the moving axis may be frozen for an instant at a time and treated as a fixed axis. This is also the second method of displaying vectors and analyzing velocities and is considered to be as economical as resolution and composition.

The LG elements of a model rotate or oscillate about their respective fixed axes and the connecting elements, called floating links, rotate with an absolute angular velocity about an instantaneous axis (IA) of velocity. The absolute instantaneous linear velocity of points on the links are proportional to the distance of the points from the IA and are perpendicular to scalars joining the points with the IA.

In this example, the absolute linear velocity of A is known and point B has a velocity sense of L. The IA of velocity may be displayed by locating the intersection of the scalars perpendicular to the directions of the velocities of A and B. At the instant in time under consideration all points in the link are tending to rotate about IV. The vector V(B), displaying magnitude, can be positioned along the sense L by the use of similar triangles, as shown in CD 6.

9.12 Instantaneous Axis Display

It should be clearly understood at this point that:
1. there is one IA of velocity for each member in a lean model,
2. there is not one common IA for all members in a lean model,
3. the IA changes position as the element moves in CD 6. The IA for CD 6 can be displayed. While the velocity vectors V(23), V(24) and V(34) for CD .6 are shown clearly in the cd figures.

9.13 Acceleration Analysis

Simply stated, acceleration in a lean model is the rate of change of linear or angular velocity. For our purposes it is convenient to classify the motion shown in CD 7 according to the kind of acceleration the model displays:

1. Acceleration zero (constant velocity)
2. Acceleration constant
3. Acceleration variable according to a simple computer display variable expressed in terms of V, L or T.

9.14 Acceleration Display Scales

The acceleration scale, designated S(ka), is defined as the linear acceleration in distance units per unit of time per unit of time, represented by 1 display unit on the monitor. If the linear acceleration of a point is 100mm /sec*sec and S(ka) is 100, a line 1 display unit long would represent a linear acceleration of 100 mm/sec*sec and would be written as:
S(ka) = 100

9.15 Linear Acceleration

Since velocity involves direction as well as rate of motion, linear accele- ration may involve a change in speed or direction as shown in CD 8. Any change in the speed of the lean model points takes place in a direction tangent to the path of the model point and is called tangential acceleration. A change in display direction as shown in CD 8, takes place normal to the display path and is called normal acceleration. Given this, acceleration may be either positive or negative. For example, if the speed is increasing, the acceleration is positive. However, if the speed is decreasing, the acceleration is negative and is called deceleration.

9.16 Angular Acceleration

As in linear acceleration, a change in either lean model speed or direction of rotation or both may be involved as shown in CD 9. Angular acceleration is understood to refer to a change in angular speed. Angular acceleration is expressed in angular units, change of speed per lean model unit of time (such as radians, degrees, or revolutions per minute). Most lean models require the use of the radian, which is the angle subtended by the arc of a circle equal in length to its radius.

9.17 Other Types of Lean Model Motion

The display of a lean model in motion is called animation. The display of a animated lean model operating under normal conditions as illustrated in Chapters 18 through 20 is called simulation. Simulation involves the acceleration of a moving model part and may vary as some function of the distance moved, velocity or time. When certain model conditions exist, definite equations may be written expressing the relation- ship among A (acceleration), L (distance), V (velocity and T (time). Three cases will be considered in Chapter 18 Gears and Cams:

1. Harmonic

2. Parabolic
3. Uniform

Often no direct relation exits among A, L, V and T that can be
conveniently expressed in the form of equations. The database for
the lean model may be obtained by observations or computations
at certain frequent intervals in an animation or simulation display.
These intervals correspond to the cycle of motion and the animation
worked out. The process of displaying simulation problems of this
type consists of approximating, by means of animation graphics, the
necessary different- iations or integrations instead of solving them
directly from equations.

9.18 Chapter Summary

This chapter introduced vector model geometry as the study of graphic
statics, model velocity and acceleration analysis. This inferred that
lean models were structures that withstand stresses. The stresses are
either pushing or pulling on each member. A pushing stress is called
compression and has a + sign. A pulling stress is called tension and has
a - sign. If the stresses are stationary, static vectors are displayed, in
motion dynamic vectors were displayed. Like lean geometry of the last
chapter, vector geometry contains powerful tools for problem solving
and each C figure in the chapter contained side by side comparisons.
When AME is used to automate this problem solution, we call it
CAVG and it is recorded as video. This represented the forces that
act in various members of a lean model. Combined with the skills
outlined in Chapter 8, stresses were presented more accurately than
the various members were sized, since in sizing we selected, from a
handbook. Using vector geometry, a model designer computed a size
and then applied a factor of safety for the lean model.

Customarily, problems in statics or dynamics were solved by
manual algebraic methods. In this chapter we assumed that you
are familiar with one or the other. With no previous background,
you will gain little from the study of how to automate it. Before
proceeding to the next further; (1) study the screen outputs shown in

Table 9.1 which compared LG and VG display methods, (2) many excellent references exist for studying the computer generated video clip. You should consult a reference source if any of these terms are unclear. We completed the study of VG by building directly on the skills learned in the last chapter.

10

Wireforms and Lean Design

The purpose of this chapter is to concentrate on one of the four major types of lean models introduced earlier in this book. A more sophisticated method of describing objects than selecting primitives or blocks as shown in Chapter 6, wireform generation models them in 3-D space. The first of the four methods, and the simplest form of lean modeling is called wireframe. A wireframe model is a skeletal description of a 3-D object. There are no surfaces in a wireframe model. It consists only of lines called wires. Because we use the AME package to display this model, the most often used pull-down menu is the model submenu DISPLAY as shown in this chapter heading.

This is the most efficient method of placing wires within the model. You can set the diameter of each wire, called density. Wireframe lean models can be used in a wide variety of applications to provide a superior method of describing and examining objects as they really exist. Some examples of applications are: 1) Enhancing visualization of 3D objects, 2) Automatically generating orthographic and auxiliary views, 3) Generating a wireframe on which to create surfaces, 4) Visual interference checking, 5) Reducing the need to create prototype models, 6) Approximating volume and mass, 7) Axonometric view construction,

8) Finding intersections of 3-D objects, and 9) Determining distance between nonintersecting objects.

10.1 Enhancing Visualization of 3-D Objects

In this section you will apply what you learned in Chapters 3, 6 and 9. Here we were introduced to the command VPOINT and VPORT. These two commands will allow you to display alternate three dimensional views of any model you have created and stored in AME. We begin the process by going to the CAD directory and loading the file airplane.scr. Please refer to Chapter 9 and C 10.1 if you do not remember how to do this. The airplane video clip was supplied on the sample video clip disk that came with CAD supplied by DLR Associates and was loaded into the central controller in Chapter 3. The sequence for doing this from the keyboard is to select an option from the main as:

Enter selection: 2
Enter name of slide file: airplane

It is difficult to tell from C 10.1 if this view of the airplane is a wireframe. To check this enter the following commands:

Command: VPORTS
Save/Restore/Delete/Join/Single/?<3>: 4
Command: VPOINT
Rotate/<View point> <0,0,1>: 1,-1,1

Your laptop should look like C 10.2. The VPOINT command gives you the last viewing position between the < > marks, you will note that airplane's viewing position was <0,1,1>, this is a plan view of the airplane. You changed it to <1,-1,1> and got four viewports. If you selected the R option of VPOINT instead of <View point> the command line on the monitor would look like:

Command: VPOINT
Rotate/<View point>:R

Enter angle in X-Y plane from X axis:30
Enter angle from X-Y plane:50

and your laptop should look like example C 10. 3.

In addition to manual typing from the keyboard, you can interactively specify a new view by manipulating an axis tripod and a compass as explained in Chapter 8.2, example (2). As you move your pointing device, the X, Y and Z axes of the tripod will animate on the screen. A small pointer inside the compass will also move simultaneous to help visualize your viewing position. The cross in the compass represents the X, Y and Z axes of the current UCS (see Chapter 3). Moving the pointer allows you to specify a view in front, back, left, right, above, or below the model as shown in Chapter 3.

10.2 Automatic View Generation

Technical orthographic views can be automatically generated using the UCS command PLAN or VPOINT<0,01> for the top or horizontal view and VPOINT<0,1,0> for the front view. VPOINT<1,0,0> will display a side view. You can try this now with the file airplane.dwg or you can load another wireframe from the directory. In CD 2, the CAD file, bracket.sld, is displayed first as a wireform, then as a dimensioned wireform and then as two technical orthographic views as shown in C 10.4.

While this is useful, may times technical views exist without a 3-D model reference. Additional Examples, will illustrate how to take a set of orthographic views and produce a lean model. Two or more orthographic views contain the XYZ filtered coordinates necessary to construct 3-D objects. While an entire chapter is devoted to this process, Chapter 14, for now all you need to know is that XYZ filters are extremely useful when converting between 2-D orthographics and 3-D axonometric views. This method of point and line specification allows you to filter or extract X, Y, and/or Z coordinate values out of existing geometry in order to edit or create new geometry. In CD bracket.

dwg, we employed point filters in a two dimensional application called orthographic views. Here's how it works.

If you pointed at a location in the top view shown in the upper left view port, then X and Y data was returned. However, if you pointed to the same point in the front view, X and Z points were returned. When you point to a location as shown in the lower right viewport, the pointing device sends a request to the display software to find that point. The software uses this information to locate a point in the database.

10.3 Using XYZ Filters to Add Lean Database

If we can use XYZ filters to locate existing display points within a database then it would make sense that we could also make the data listing longer by adding additional points. These points are starting and ending points for wires. Wires are added in the upper left viewport, the UCS is changed in the lower left so that even more points may be added in the upper right viewport. Once all the points have been added to the data listing, then the HIDE command can be used to view the wireframe without hidden lines. In this manner, old existing databases can be used to create new models.

10.4 Using a Wireframe to Create Surfaces

The combination of XYZ filters and the command HATCH can be used to create detailed surfaces as in CD 1, 2 and 3, or reference surfaces. If detailed surfaces are required, refer to C figure 10.8. Here a sample floorplan was taken from the CAD directory and displayed in the upper left viewport. The UCS was changed and the menu item EXTRUDE was used to create a wireframe model. In the lower left viewport the second story was added using another floorplan sample and extrusion. In the upper right viewport the UCS was changed and the windows were then added. Finally in the lower right viewport, the UCS was changed to create the roof wires and hatch between the wires.

You can repeat this same process because all the pieces exist in the main directory. The pieces do not exist either as prior CAD directory files or as building primitives. In this case you can construct the geometry as shown in Chapter 8. This will create the software database.

Once this is done, then XYZ filters can be used to create the top view (X,Z) in the upper left viewport, the front view (X,Y) in the lower left viewport and the side view (Z,Y) in the lower right viewport. In this example, the back surface of the lean model was hatched to indicate that it is a reference surface. Dimensions are displayed relative to these types of surfaces called datums. Datums always appear as edges in dimensioned views, here the arrows point to them in the upper left and lower right viewports.

If XYZ filters are used, the reference surface can be made to appear to move in 3-D space as shown in the example C figures. The data list X, Y or Z is multiplied by a -1. This does not change the values of the list, but does change its location relative to the display axis. In the CD images, the wireframe appears to roll around the X axis, while in CD 3 it appears to pitch around the Y axis, and in CD 4 it appears to yaw around the Z axis. Remember the wireframe is not moving. Data is being multiplied by -1, this is called translation. By creating one surface on the wireframe it is easier to visualize the translation about the display axis.

10.5 Visual Interference Checking

Wireframes are useful because they are quick to generate and have many applications in the field of lean modeling. One of these is visual interference and clearance checking as shown in the video clips. In a design office you will have many 2-D and 3-D files created for a single product design like a robot. What is the best way to see how pieces fit together?

Remember XYZ filters? With these we can combine 2-D files with 3-D models. If the 3-D models need to be machined during production, why not call the 2-D tools and place them near or inside the 3-D models to check clearance or interference. In the upper left viewport a drill has clearance through the model. After drilling we may want to ream the hole. We can check for clearance as shown in the upper right viewport. Some tooling requires a clearance and then an interference in order to operate. In the lower left viewport a spotface cutter illustrates this. In the lower right viewport, a counterbore also requires both in order to operate.

10.6 Using Wireframes as Prototype Models

A prototype model is the original model or the model that all others are patterned after. Some chapters in this text lend themselves to prototype models others do not. For example, Chapter 18, Lean Gear and Cam Models uses prototypes. In this case gears are ideal because every gear tooth is based upon a known pattern, while at the same time cams are not. Cams are designed to solve a unique motion problem and are not good candidates for prototype modeling.

With this in mind, video clips contain CD images, and were produced for use in Chapter 18. In the upper right viewport a worm gear and driver known as the worm are shown as a wireframe prototype. Using XYZ filters in the upper left view- port, two working views are displayed showing the prototype variables: outside diameter (OD), throat diameter (TD), pitch diameter (PD), center distance (CD), face width (FW), face radius (FR) and rim radius (RR). All of these variables are used in Chapter 18 to produce different sized gear teeth as shown. The worm variables are shown in the lower right viewport and are similar to those for a screw thread shown in Chapter 19. The thread profiles are shown and are sized to match the gear teeth shown.

10.7 Approximating Volume and Mass

The volume calculation of wireframes is an approximation as shown in the CD images, while the mass (weight) of a wireframe is zero. In order to get a mass we need to assign a material property to the wireframe with the command SOLMAT. If we assign something like nylon instead of brass for the objects pictured in Table 10.3, the mass listed in the lower view port will be very different. The mass properties for a wireframe model once the material is known are: Mass/Bounding box/Centroid/Moment of inertia/Products of inertia/Radii of gyration/ Principal Moments and are requested by typing M, B, C, M, P, R or PM.

The **SOLMAT** command sets the default material from a material in your wireframe file. If none is listed the command will ask for input.

Command: SOLMAT

You will find other material files in the CAD directory for all of the common manufacturing materials used in Chapter 20.

10.7 Axonometric Views for Wireframes

The most common axonometric views are shown in the video clips. They are isometric shown in the upper left viewport, dimetric shown in the upper right viewport and trimetric shown in the lower left viewport. Obliques shown in the lower right viewport are either cabinet or cavalier depending upon the depth scale. You present axonmetric views with the VPOINT command. The isometric has R values of 30,30, the dimetric has R values of 15,15, the trimetric has R values of 15,30 and the oblique has R values of 0,30.

10.8 Displaying Intersections of Wireframes

The CD images illustrates several intersections. In the upper left viewport an intersection is called for between cylinder and box. This is done with the **SOLINT** command as shown. Intersections can be between wireframes and voids (holes) as shown in the upper right viewport. Even different types of holes can be displayed, use:

Command: HOLE
Csink/<CBore>:

to display either a counter sink or a counter bore.

10.9 Finding Distance Between Wireframes

You can find the distance between wireframes by using the **STATUS** command. You need to practice the use this to determine the distance between wireframes. The model space limits and uses are listed in the lower viewport for each wireframe shown in the upper viewport. You must do some simple subtraction to determine the display space and UCS space between each wireframe as shown.

10.10 Chapter Summary

The purpose of this chapter was to concentrate on one of the four major types of lean models introduced earlier in this book. A more sophisticated method of describing objects than selecting primitives or blocks as shown in Chapter 6, wireform generation models them in 3-D space. The first of the four methods, and the simplest form of lean modeling was called wireframe. A wireframe model was a skeletal description of a 3-D object. There are no surfaces in a wireframe model. It consists only of lines called wires. Because we use the AME package to display this model, the most often used pull-down menu is the model submenu **DISPLAY**.

This was the most efficient method of placing wires within the model. You can set the diameter of each wire, called density. Wireframe lean models wee used through out the chapter in a wide variety of applications to provide a superior method of describing and examining objects as they really exist. Some examples of applications were: 1) Enhancing visualization, 2) Automatically generating orthographic views, 3) Generating wireframe surfaces, 4) Visual interference checking, 5) Create prototype models, 6) Approximating volume and mass, 7) Axonometric view construction, 8) Displaying intersections of wireframes, and 9) Determining distance between nonintersecting objects.

11

3D Surface Modeling

The purpose of this chapter is to concentrate on the second of the four major types of lean models introduced earlier in this book. A more sophisticated method of describing objects than wireframes as shown in Chapter 10, 3D surface models use planes in 3-D space. The second of the four methods, and still a fairly simple form of lean modeling is called 3D surface modeling. A surface model is an infinitely thin shell that corresponds to the shape of the object being modeled. This shell consists of a combination of flat and curved surfaces or adjoining surface elements called patches depending upon the type of surface model used. The main types of surface models are:

1. Extrusions,
2. Ruled surfaces,
3. 3D faces,
4. 3D meshes,
5. Smoothed meshes,
6. Tabulated surfaces,
7. Revolved surfaces and
8. Edge surfaces.

11.1 Wireframe VS Surface Modeling

There are no surfaces in a wireframe model and there are no solid insides to a surface model as shown in the CD images downloaded for Chapter 11. Wireframes consists only of lines called wires and 3D surface models contain only flat planes or curved patches. The best way to demonstrate this is to give a person a bottle of glue and a handful of soda straws then ask them to build a 3D object, the result will be a wireframe. Give that same person some card board a pair of scissors and roll of duck tape and the result is a surface model.

Because we use the AME package to display a surface model, the most often used pull-down menu is the submenu 3D surfaces as shown in the chapter heading. This is the most efficient method of placing the flat planes and the curved patches within the model. You can set the mesh size of each patch, called resolution. Like wireframe lean models, 3D surface models can be used in a wide variety of applications to provide a superior method of describing and examining objects as they really exist. Some examples of applications are: 1) Assigning vertex data information to 3D objects, 2) Automatically generating finite element views, 3) Generating a 3D surface to create solids, 4) Surface measurement for tolerance checking, 5) Determining the MMC and minimum material conditions, and 6) Approximating the mass, and determining the surface contact between two tangent objects.

11.2 Constructing Surface Models

In this section you will apply what you learned in Chapters 6, 8 and 10. Here we were introduced to the command VPOINT and VPORT. These two commands will allow you to display alternate three-dimensional views of any model you have created and stored in C figures 11.1 through 11.4. We begin the process by reviewing the ELEV command. In C figures 7, 8 and 9, are example shapes displayed as a sample floor plan and wireform parts. Next the ELEV command was used to set a new current elevation of .5 as shown in C figures10, 11, and 12. The VPOINT command yielded the result shown in C figures 12, 13 and 14. In C 15, 16, and 17, a new viewport and layer were called for with the following wireframe images displayed:

Next the RULESURF menu item was selected and the surfaces were ruled as shown in C figures 4, 9, 20, 41 and 43. The database listing for the wireframe was used along with the 3D FACE menu item to create C figures 5, 7, 10, 1123, and 37. Next four holes were placed in the extruded layer shown in C figures 17, 35, 36, and 37. These CD images demonstrates how holes and cylinders can be extruded.

The menu item EDGESURF was used in C figures 25, 26, 27, 28, 29, 30 and 31 to illustrate 3D meshes. The layers 1 (extrude) and layer 2 (ruled) were turned on to complete C figures 33, 37, 38, and 42.. One or more layers 1, 2, 3 (faces) or 4 (meshes) can be combined to form the 3D surface model displays shown in these chapter CDs.

11.3 3D Models From Extruded Surfaces Only

Using the downloaded CD images again we can extrude the holes and add additional extruded sections as shown in C figures 8, 9, 20, 23 and 43. In these CDs, we can extrude either holes or shafts.

11.4 3D Models From Ruled Surfaces Only

Referring again to the download, we can create ruled surface pin connectors as shown in C 4, 18, 34, 39 and 43.

11.5 3D Face Models From Wireframes

In Chapter 10 we saw how orthographic views were automatically generated using the UCS command PLAN or VPOINT<0,01> for the top or horizontal view and VPOINT<0,1,0> for the front view. VPOINT<1,0,0> displayed a side view. We will use this technique to convert a wireframe to a 3D face model.

This is very useful, may times wireframe models exist without a 3-D face model reference. The next examples illustrate how to take a set of orthographic views or wireframes and produce a 3D face lean model. Two or more adjacent orthographic views contain the XYZ filtered coordinates necessary to construct wireframe 3-D objects. While an entire chapter is devoted to this process, Chapter 14, for now all you need to know is that XYZ filters are extremely useful when converting between wireframe and 3-D face models. This method of point and line specification allows you to filter or extract X, Y, and/or Z coordinate values out of existing geometry in order to edit or create new geometry. In C 11.1 we employed point filters in a three dimensional application called wireframe. Here's how it works.

C 11.1 was displayed from the database shown in Chapter 10. We need four points for the command 3DFACE to work. So if we wanted to create a 3D face composed of wireframe points 1,2,3,4,5,6,1 as shown in the upper left hand viewport in C 11.1 we really need to think of that face as two commands. The first being 1,2,3,4,1. If we enter the data as:

Command: 3DFACE
First point:0.0,0.0,0.0
Second point:2.0,0.0,0.0
Third point:2.0,1.0,0.0
Fourth point:1.0,1.0,0.0,i
and if you enter the second command as:
First point:1.0,1.0,0.0
Second point:1.0,1.5,0.0
Third point:0.0,1.5,0.0
Fourth point:0.0,0.0,0.0,i

we get CD 5. Using this technique, we can join two more faces as shown in the remaining CDs 6 and 7.

11.6 Using XYZ Filters to Edit Wireframe Database

If we can use XYZ filters to locate and form existing display planes called faces within a database then it would make sense that we could also make the data listing longer by adding or subtracting additional points to form other faces as shown in Chapter 10. These facess are starting and ending edges for planess. Planess are added in the other viewports. Once all the planes have been added to the data listing shown in Chapter 10, then the HIDE command can be used to view the wireframe without hidden lines. In this manner, old existing wireframe databases can be used to create new 3D face surface models as shown in CD 5, 6, 7, 10, 11, 12, 13, 14, 15, 16,17, and 18.

11.7 Using a 3D Mesh to Create Surfaces

In the downloads, CD 28, we see how a five by four grid can be used to create a mesh surface. All twenty pieces of vertex data must be read from a file, entered by a pointing device or typed into the command stream 3DMESH. You may edit data by pointing to the screen and changing the location of a vertex. Once the raw data has been entered and edited, it can be smoothed as shown in CD 25. This is done by the SETVAR command which asks for a SURFTYPE. A quadratic smoothing shown in CD 29 is entered or selected as 5. A cubic smoothing shown in CD 30 is 6 and the bezier in CD 31 is 7.

11.8 Tabulated and Revolved Surfaces From Wireframes

Wireframes are useful because they are quick to generate and have many applications in the field of lean modeling. One of these is visual interference and clearance checking as shown in Chapter 10. In a design office you will have many wireframe files created for a single product design like a robot. What is the best way to see how pieces fit together?

Remember XYZ filters? With these we can combine wireframe files with 3-D faces. If the 3-D facess need to be machined during production, why not call the 3-D tools and place them near or inside the 3-D models to check clearance or interference as shown in Chapter 10. CD 34 shows a wireform has clearance through the model at point P2. We can use the TABSURF menu item to create a surface model as shown in CD 39.

After we check for clearance as shown in CD 34, we can revolve a surface as shown in CD 40. Some tooling requires a clearance and then an interference in order to operate as shown in Chapter 10, for this we can rule those surfaces as shown in CD 43. To complete the 3D surface model from the wireframe, EDGESURF is selected from the menu and is used in CD s 37, 38, and 41.

11.9 Using 3D Surface Model Prototypes

A prototype model is the original model or the model that all others are patterned after. Some chapters in this text lend themselves to prototype models others do not. For example, Chapter 18, Lean Gear and Cam Models uses prototypes and wireframe models are ideal, because every gear tooth is based upon a known pattern. But at the same time cams are not suited for wireframes, they are better suited for surface modeling. Cams are designed to solve a unique motion problem and are not good candidates for wireform modeling.

With this in mind, Table 11.4 was produced for use in later chapters. In CD 42 an extruded and 3DFACE surface fan model is shown. This is the prototype for other fan blades. Using XYZ filters in the other CD s, working views are displayed showing the prototype variables: fan rotation, offset, offset circle, base circle, working curve, theoretical curve, and blade locations. All of these variables are used to produce different shaped fan blade.

11.10 Approximating The Mass Of A Model

A surface model has a volume and can be calculated, but has no weight. The volume calculation of a wireframe is an approximation because we can assign surfaces between the wires as shown in Chapter 10. While the mass (weight) of a surface model is zero, we get a mass because we assign a material property to the surface model with the command SOLMAT. If we assign something like nylon instead of brass, the mass will be very different. The mass properties for a surface model, once the material is known, are: Mass, Bounding, Box, Centroid, Moment of inertia, Products of inertia, Radii of gyration, Principal Moments and are requested by typing M, B, C, M, P, R or PM.

The SOLMAT command sets the default material from a material in your surface file. If none is listed the command will ask:

Command: SOLMAT
Change/Edit/LIst/LOad/New/Remove/SAve/SEt/<eXit>:N
Enter material name: brass

You will find other material files in the CAD directory for all of the common manufacturing materials used in Chapter 20.

11.11 Chapter Summary

The purpose of this chapter was to concentrate on the second of the four major types of lean models introduced earlier in this book. A more sophisticated method of describing objects than wireframes as shown in Chapter 10, 3D surface models used planes in 3-D space. The second of the four methods, and still a fairly simple form of lean modeling was called 3D surface modeling. A surface model was an infinitely thin shell that corresponded to the shape of the object being modeled. This shell consisted of a combination of flat and curved surfaces or adjoining surface elements called patches depending upon the type of surface model used. The main types of surface models were:

1. Extrusions,
2. Ruled surfaces,
3. 3D faces,
4. 3D meshes,
5. Smoothed meshes,
6. Tabulated surfaces,
7. Revolved surfaces and
8. Edge surface.

12

Solid Modeling

The purpose of this chapter is to concentrate on the third of the four major types of lean models introduced earlier in this book. A more sophisticated method of describing objects than wireframes or surfaces as shown in Chapters 10 and 11, solid modeling use solids in 3-D space. The third of the four methods is called solid modeling. This is a slightly more difficult form of lean modeling. A solid model is a completely filled surface or wireframe corresponding to the shape of the object being modeled. A solid model may consists of a combination of sections called regions or solid composites depending upon the type of lean model created. The main types of solid models are:

1. Extruded regions or solid primitives,
2. Revolved regions or composite solids,
3. Boolean regions or solids,
   - A. **SOLINT** (intersections),
   - B. **SOLSUB** (subtractions),
   - C. **SOLUNION** (unions),
4. Shaped regions or solids,
5. Smoothed regions or solids and
6. Topology regions or solids.

This chapter deals with solids and Chapter 13 deals with regions.

12.1 Wireframe, Surface or Solids Modeling

There are no surfaces in a wireframe model and there are no solid insides to a surface model. Wireframes consists only of lines called wires and 3D surface models contain only flat planes or curved patches. The best way to demonstrate this is to give a person a bottle of glue and a handful of soda straws and ask them to build a 3D object, the result will be a wireframe; some card board, a pair of scissors and roll of duct tape and the result is a surface model: a bar of soap and a carving knife and the result is a solid model.

Because we use the AME package to display solids, the most often used command menus are the SOL --- side menus, one of the seven is shown in the chapter heading. These are the most efficient methods of designing a solid model. A solid model is an unambiguous, complete representation of the shape of a physical object. It can be analyzed to give accurate information about properties (**SOLMASSP**) or surface area (**SOLAREA**). Some examples of applications are:

1) Accurate data for dimensioning 3D objects, 2) Automatic generation of section and profile models, 3) Generating lean tolerances, 4) Surface measurement for manufacturing (Chapter 20), 5) Determining MMC and mmc for threads and 6) Accurate analysis of the mass and determining the surface contact between two tangent objects.

12.2 Constructing Solid Models

In this section you will apply what you learned in Chapters 10 and 11. Here we were introduced to the command **VPOINT** and **VPORT**. These two commands will allow you to display alternate three-dimensional views of any model you have created and stored in AME. We begin the process by reviewing the solid primitives available in AME: 1) **SOLBOX**, creates a solid box or cube, 2) **SOLCONE**, creates a cone, 3) **SOLCYL**, creates a cylinder, 4) **SOLEXT**, extruded solid, 5) **SOLIDIFY**, solids from objects or regions, 6) SOLREV, solid from revolved region, 7) **SOLSPHERE**, solid sphere, 8) **SOLTORUS**, solid donut, and 9) **SOLWEDGE**, a solid wedge.

Solid primitives are the basic building blocks that make up complex solid models. Primitive commands, as shown in Table 12.1, can create solids of predefined shapes or user defined shapes as shown. Examples of predefined solids as the **SOLBOX, SOLCYL, SOLSHAFT** and **SOLWEDGE** shown here. An example of a user defined primitive is *DRILL C.* There are no regions used in Table 12.1 because you create regions from 2D entities with the **SOLIDIFY** command or from sections of solids as:

CD Images of Solid Primitives

Command	*Function*
SOLBOX	*Creates a solid box or cube*
SOLCONE	*Creates a solid cone*
SOLCYL	*Creates a solid cylinder*
SOLEXT	*Creates extruded sections*
SOLIDIFY	*Creates solids from objects*
SOLREV	*Creates solids by revolving*
SOLSPHERE	*Creates solid spheres*
SOLTORUS	*Creates solid torus*

12.3 Solid Models From SOL primitives

Limited to box, cone, cylinder, sphere, torus and wedge; solids are some what limited to Table 12.1 Even here a cone and a cylinder, a sphere and a cylinder were joined to form the last two primitive examples. When two or more primitives are joined or edited, the result is called a composite model. The commands for composite models are: 1) **SOLINT**, creates solids that are in the intersection of primitives, 2) **SOLSEP**, breaks apart composites, 3) **SOLSUB**, creates composites that are subtractions, and 4) **SOLUNION**, a new solid from two or more composites.

In C Figure 12.2, upper left view port is composed of a solid primitive command **SOLBOX** and solid user defined primitive *DRILL*, so this an example of a solid primitive. The upper right view port contains the solid primitive commands **SOLCYL** and **SOLSHAFT** and user defined primitive *DRILL* plus the two commands **SOLSUB** and **SOLINT**, this makes this a composite model.

Solid Modeling Editing Commands are: 1) **SOLCHAM**, bevels edges, 2) **SOLFILL**, rounds edges, 3) **SOLCUT**, cuts solids with a plane, 4) **SOLCHP**, edits a solid even if it is part of a composite model, and 5) **SOLMOVE**, moves or rotates a solid.

12.4 Composite Solid Models

Using the editing and the composite commands we can create more realistic models as shown in C figure 12.3. Here, both editing and composite commands were used. For example, the upper right viewport contains a union between a solid primitive cylinder, box, box, and cylinder. A user defined primitive called *DRILL* displays the holes and the slot was displayed by the **SOLSUB** command. In the upper left view port, part 1 is composed of box + wedge (union of primitives), cylinder + box (three of these), and *DRILL*. In the lower left view port, part 3 is composed of cylinder+box+cylinder and *DRILL*. In the lower right view port, part 11 is composed of cylinder and *DRILL*.

12.5 BOOLEAN Models From Primitives

Composite models are constructed from solid primitives using solunion (+), solsub (-), solint (I), and solsep (<>). These operations (+, -, I, <>) are called Boolean operations and can be tree diagramed for an composite model by using the **SOLLIST** command as:

Command:SOLLIST
Edge/Face/Tree/<object>: TREE
Select objects:
Select objects:
If we select object 2 from Table 12.1 the following listing is given:
Object type = union   Handle = CYL + BOX + BOX
Component handles: 2a + 2b + 2c Area not computed

Material = Mild Steel Representation = Primitive Render type = CSG (Constructive Solid Geometry) Object type = intersection Handle = Box I CYL Component handles: 2c I 2d
Area not computed Material = Mild Steel Representation = Primitive Render type = CSG Object type = subtraction Handle = DRILL Component handles: 2a - DRILL, 2c - DRILL, 2d - DRILL Area not computed Material = Mild Steel Representation = SOL variable DRILL Render type = CSG

The actual tree diagram looks like the CD image.

12.6 Shaped Solid Models

If we can use *DRILL.C* for drilling holes in solids, then maybe we can use other C type routines. AME has the following prestored.

1. TUTOR.C ... teaches basic function programming which includes: Functions -- SOLPIPE, SOLTOPOL, SOLTRACE, SOLNORM, SOLTAN, SOLCLASS, SOLUCSPD.
2. ASM.C .... provides feature based assembly of solids includes: Functions -- SOLCONTACT, SOLALIGN.
3. DRILL.C .... drills holes in solids contains: Function SOLDRILL.
4. DESIGN.C..... models basic machine parts includes Functions: SOLSHAFT, SOLWHEEL, SOLGEAR, SOLBEAR, SOLBOLT, and SOLNUT.
5. LAYOUT.C..... lays out sheet blanking model. Functions GEAR, SHEET, LAYOUT.
6. SYMMETRY.C. determines symmetry of models. Function SYMMETRY.
7. OFFSOL.C...... creates offset polylines for smoothed models. Functions -- SOLOFF, SOLMAC.

In Table 12.2 we see how to apply these in the display of a shaped solid model. Let's begin with part 39. **SOLGEAR** was used to display the basic sizes of the ring, outside and inside diameter as shown. Next a **SHEET** command large enough to contain the required eight gears was used as shown, and finally the **LAYOUT** command was used to produce the amount needed. The whole purpose of shaped solid models is to take basic starter pieces and shape them into the necessary shapes to meet the demands of the lean model being designed.

Not all shaped models begin with prestored elements, in Table 12.2, parts 1, through 9, 11, 15, 16, 17, 30, 31, 39, 43, and 45 were all constructed from prestored functions, commands or variables. The rest of the parts were constructed in the normal manner already shown in this chapter.

12.7 Smoothed Solid Models

In the chapter heading model, we see how a solid model can be become more realistic by a smoothing routine called **SOLTOPOL** used with SOLMESH. Solids can not be shaded or smoothed unless they are assigned a polymesh surface. If solids are to be represented as cast metals, then they will not have sharp corners because these are difficult to cast and will result in weak points. The cast surfaces were covered with the **SOLMESH** command. Just like last chapter, there are three types to choose from.

12.8 Topology of Solid Models

The **SOLTOPOL** command lets you query topological relationships for a solid model. It counts the total number of faces, edges, and vertices associated with the selected solid. Solid models are established on the basis of complete lean information and sound topological relationships. The topological elements of a solid consist of bodies, faces, edges, and vertices The solid's topology deals with the structure and adjacency of topological elements, such as finding all edges surrounding a given vertex, finding the faces sharing the given edge, finding all edges associated with a given face, and so on.

In the chapter heading, **Facelist** and **Edgelist** were the data structures that support this application. Once these were found then **SOLMESH** could cover these surfaces. The Render pull-down menu was then used to select the surface finishes and shading shown. Please review Chapter 6 for how to render lean models.

12.9 Using Solid Models as Prototype Models

A prototype model is the original model or the model that all others are patterned after. Some chapters in this book lend themselves to prototype models others do not. For example, Chapter 18, Lean Gear and Cam Models uses prototypes and solid models are ideal, because every gear tooth is based upon a known pattern. But at the same time cams are not, they are better suited for surface modeling. The chapter heading is an example of a prototype solid model it contains all the modeling techniques, including regional sections and was produced for use in Chapter 18.

12.10 Determining The Mass Of A Solid Model

A solid model has a mass and can be calculated. The volume calculation of a solid is accurate because we can find the topology of a solid. While the mass (weight) of a solid model is known, we get a mass because we assign a material property to the solid model with the command **SOLMAT**. If we assign something like aluminum instead of cast iron for Table 12.2, the mass will be very different. The mass properties for a solid model, once the topology and material is known, are: Mass, Bounding Box, Centroid, Moment of inertia, Product of inertia, Radii of gyration, Principal Moments are requested by typing M, B, C, M, P, R or PM.

The SOLMAT command sets the default material from a material in your surface file. If none is listed the command will ask:

Command: SOLMAT
Change/Edit/LIst/LOad/New/Remove/SAve/SEt/<eXit>:N
Enter material name: Mild Steel

You will find other material files in the DLR Associates CAD directory for all of the common manufacturing materials used in Chapter 20.

12.11 Obtaining Data for Dimensioning

In the chapter downloaded CDs the topology of a model can determine all the necessary data to accurately dimension a lean model. The steps involved in this process are discussed in detail in Chapter 15, for now C 12.3 just illustrates how the various parts of the solid model are presented in working view relationship because dimensions are shown on orthographic views most often. As seen in Chapter 10, orthographic views are automatically generated with filtered XYZ data. More information may be obtained by reference to Chapter 14 (Lean Dimensioning and XYZ Filters).

12.12 Automatic Generation of Section Views

The side view, in C 12.3, represents a broken out section. The steps involved in this process are discussed in detail in Chapter 13, for now C 12.3 shows how regions can be produced from a solid model or from an XYZ filtered view. Regional sections are very helpful because often times only an orthographic view of an object exists within a directory. This is enough, first pass a section cutting plane through the orthographic view, this will produce one or more sections in the adjacent view. Once this has been done, select this cross-section and use the **SOLIDIFY** command to produce a region. This region can then be used to produce the required solid model as shown in Chapter 13.

12.13 Generating Lean Tolerances

In C 12.3, the topology of a model can determine all the necessary data to accurately dimension and tolerance a lean model. The steps involved in this process are discussed in detail in Chapter 16, for now it just illustrates how the various parts of the solid model are presented in working view relationship because dimensions and tolerances are shown on orthographic views most often. As seen in Chapter 10, orthographic views are automatically generated with filtered XYZ data. More information may be obtained by reference to Chapter 14 (Lean Dimensioning and XYZ Filters).

12.14 Surface Measurement for Manufacturing

Surface measurement for manufacturing is used in both lean tolerancing and MMC. C12.3 was created to illustrate the three states of lean tolerancing. The smallest measurement is the least material allowed and is known as minimum material condition (mmc). This is the smallest size that can be manufactured and still meet the topological model. The largest measurement is the MMC (maximum material condition) allowed. This is the largest size that can be manufactured and still meet the topological model requirements. The surface is the basic design size and is the ideal topological model measurement. See Chapter 20 for more details on surfaces.

12.15 Determining MMC and mmc for Threads

All manufactured parts including threads and fasteners must meet MMC and mmc conditions. This is particularly important when you have mating parts such as holes and pins or bolts and nuts. When both haves are free to move for assembly we refer to these as floating fasteners. If we using stud bolts welded in place, we refer to that as a fixed fastener. We can have any possible combination of manufactured sizes for either the bolts or the nuts shown in Table 12.1. Let's take the two worst conditions; where we have a very small (mmc) bolt and a very large nut opening (mmc). The fit when assembled will be loose. The opposite case, large bolt small nut opening will result in a tight fit. It is important in Table 12.1 that all manufactured items fit according to there class of fit. See Chapter 19 for more information.

12.16 Determining Surface Contact for Solids

One of the many uses for topological solid models is the determination of surface contact to measure friction, wear support points and bearing surfaces. The chapter heading is a topology model for several different types of bearings: self-aligning, single row deep groove, angular contact, double row deep groove, spherical roller, and tapered roller. Because this is also a prototype model it can be used to compare one choice with another for a certain application within manufacturing. See Chapter 20 for more information and additional examples.

12.17 Chapter Summary

The purpose of this chapter was to concentrate on the third of the four major types of lean models introduced earlier in this book. A more sophisticated method of describing objects than wireframes or surfaces as shown in Chapters 10 and 11, solid modeling used solids in 3-D space. The third of the four methods was called solid modeling. This was a slightly more difficult form of lean modeling. A solid model was a completely filled surface or wireframe corresponding to the shape of the object being modeled. A solid model may consists of a combination of sections called regions or solid composites depending upon the type of lean model created. The main types of solid models were:

1. Extruded regions or solid primitives,
2. Revolved regions or composite solids,
3. Boolean regions or solids,
   A. **SOLINT** (intersections),
   B. **SOLSUB** (subtractions),
   C. **SOLUNION** (unions),
4. Shaped regions or solids,
5. Smoothed regions or solids and
6. Topology regions or solids.

13

Section and Profile Models

The purpose of this chapter is to concentrate on the fourth major type of lean model introduced earlier in this book. A more detailed method of describing objects than wireframe, surface or solids modeling as shown in Chapters 10, 11 and 12, section and profile models use cross sections of solids in 3-D space called regions. The fourth of the four methods is called regional modeling. This is a slightly more difficult form of lean modeling to visualize in the mind's eye. A region model is a cross section of a wireframe, surface or solid corresponding to the shape of the object being modeled. A solid model may consist of a combination of regions, but a region model can not have a combination of solids. A region is unique to its cross sectional shape.

13.1 Types of Regional Models

The main types of regional models are:
1. Extruded regions,
   A. SOLCHP (edits a region primitive)
   B. SOLCUT (region cutting plane CP)
   C. SOLEXT (regional ELEV command)

   2.   Revolved regions,
- A. SOLREV (rotates a region)
- B. SOLMOVE (rotates a region or solid)

   3.   Boolean regions,
- A. SOLINT (intersections),
- B. SOLSUB (subtractions),
- C. SOLUNION (unions), and
- D. SOLSEP (separations).

   4.   Shaped regions and
- A. SOLSECT (cross section)
- B. SOLFEAT (region from a face or edge)
- C. SOLPROF (region from a profile view)
- D. SOLCHP (edits a region primitive)

   5.   Topology regions.
- A. SOLMESH (places a polymesh over the region or solid)
- B. SOLPURGE (removes data after utility)
   - 1. SOLIN (imports assembly files)
   - 2. SOLOUT (exports files)
   - 3. SOLMAT (material assignment)
   - 4. SOLVAR (sets AME variables)
- C. SOLUCS (aligns the UCS to the face)

This chapter deals with regions and Chapter 12 deals with solids.

13.2 Wireframe, Surface, Regional Modeling

There are no surfaces in a wireframe model and there are no solid insides to a surface model, while the solid is filled with material. Wireframes consists only of lines called wires, surface models contain only flat planes or curved patches, solid and regional models contain manufacturing materials. The best way to demonstrate all four model types this is to give a person a bottle of glue and a handful of soda straws and ask them to build a 3D object, the result will be a wireframe; some card board, a pair of scissors and roll of duct tape and the result is a surface model, a bar of soap and a carving knife and the result is a solid model. Give the same person a mold cavity and something to pour into it that will harden and the result is a regional model.

Because we use the DLR Associates AME package to display regional solids, the most often used command menus are the SOL side menus, one of the seven is shown in C 13.1. These are the most efficient methods of displaying a regional model. A regional model, like a solid model is also an unambiguous, complete representation of the shape of a physical object, because its mold cavity is its cross section. Just like solid models regions can be analyzed to give accurate information about properties (SOLMASSP) or surface area (SOLAREA). Some examples of applications are:

1. Forging Dies,
2. Foundry Patterns and Molds,
3. Stamping Dies,
4. Extrusion Molding,
5. Cold Forming,
6. Deep drawing and Spinning,
7. Roll Forming, and
8. High-velocity forming.

13.3 Constructing Region Models

In this section you will complete what you started to learn in Chapters 10, 11 and 12. Here we were introduced to the command VPOINT and VPORT. These two commands will allow you to display alternate three-dimensional views of any regional model you have created and stored in AME. We begin the process by loading the regional modeler software section of AME:

Command: (xload region)

The region modeler is part of the Auto Development System (ADS) software package. Regions are closed 2D areas from orthographic views or cross sections of solid models produced by a cutting plane (CP). C 13.1 is an example of this. The most often used regional models are:

1. Extruded, shown in the upper left viewport,

2. Revolved, shown in the upper right view port,
3. Boolean, shown in the lower left view port, and
4. Shaped, shown in the lower right view port.

13.4 Regional Models From SOLEXT Only

Any of the regions shown in C 13.1 can be extruded by using:

Command: SOLEXT
Select regions, polylines and circles for extrusion:
Select objects:
Height of extrusion: 1
Extrusion taper angle <0.>: 15.

13.5 Regional Models from SOLREV

Using the polyline region shown in the upper right view port in C 13.1, and the following:

Command: SOLREV
Select region, polyline or circle for revolution:
Select objects:
Axis of revolution - Entity/X/Y/<Start point of axis>:
End point of axis:
Angle of revolution<full circle>: 270.
Revolve individual loops to different angles <no>: N

we get the display shown in the upper right view port of C13.1. This is a rather simple case. C13.1 is also a more typical revolved model using Boolean regions.

13.6 Regional Models From Booleans

Composite models are constructed from regional primitives using solunion (+), solsub (-), solint (I), and solsep (<>). These operations (+, -, I, <>) are called Boolean operations and can be tree diagramed for an composite model by using the **SOLLIST** command as:

Command:SOLLIST
Edge/Face/Tree/<object>: TREE

If we select a region from C13.2, the following listing is given:

Region type = union Handle = CYL + BOX + BOX + CYL + BOX + CYL   Component handles: 3a + 3b + 3c + 3d + 3e + 3f

Area not computed   Material = Mild Steel
Representation = Primitive
Render type = LSG (Lean Solid Geometry)
Region type = intersection Handle = Box I CYL, Box I CYL
Component handles: 3a I 3b, 3c I 3d, 3e I 3f
Area not computed   Material = Mild Steel
Representation = Primitive Render type = LSG
Region type = subtraction Handle = SOLDRILL
Component handles: 3a - SOLDRILL, 3d - SOLDRILL, 3f - SOLDRILL
Area not computed   Material = Mild Steel
Representation = SOL variable DRILL   Render type = LSG
The result is shown in the lower left view port in C13.1. A more typical Boolean model is shown in C13.2, here only two primitives were used CYL and SOLDRILL.

13.7 Regional Shaped Models

If we can use SOLDRILL for drilling holes in regional solids, then maybe we can use other C type routines. AME has the following prestored.

1. TUTOR.C ...teaches basic function programming includes:

Functions -- SOLPIPE, SOLTOPOL, SOLTRACE, SOLNORM, SOLTAN, SOLCLASS, SOLUCSPD,
2. ASM.C provides feature based assembly of regions includes: Functions -- SOLCONTACT, SOLALIGN
3. DRILL.C .... drills holes in solids contains: Function -- SOLDRILL
4. DESIGN.C..... models basic machine parts includes Function -- SOLSHAFT, SOLWHEEL, SOLGEAR, SOLBEAR, SOLBOLT, SOLNUT
5. LAYOUT.C..... lays out sheet blanking regional models Functions -- GEAR, SHEET, LAYOUT
6. SYMMETRY.C. determines symmetry of regional models Function -- SYMMETRY
7. OFFSOL.C...... creates offset polylines for topological regions Functions -- SOLOFF, SOLMAC

In C13.2, we see how to apply these in the display of a typical regional shaped model.

13.8 Topology of Regional Models

In C 13.2 we see how a region model can be become more realistic by a smoothing routine called SOLTOPOL used with SOLMESH. Regional models can not be shaded or smoothed unless they are assigned a polymesh surface. If models are to be represented as forged metals, then they will not have sharp corners because these are difficult to draft and will result in weak points. In C 13.2 the forged surfaces were covered with the SOLMESH command, like last chapter, there are three types to choose from.

The SOLTOPOL command lets you query topological relationships for a regional model. It counts the total number of faces, edges, and vertices associated with the selected model. Regional or solid models are established on the basis of complete lean information and sound topological relationships. The topological elements of a lean model consist of bodies, faces, edges, and vertices The model's topology deals with the structure and adjacency of topological elements, such as finding all edges surrounding a given vertex, finding the faces sharing the given edge, finding all edges associated with a given face, and so on.

In C 12.2 and 13.2 Facelist and Edgelist were the data structures that support this application. Once these were found then SOLMESH could cover these surfaces.

13.9 Using Regional Models as Prototypes

A prototype model is the original model or the model that all others are patterned after. Some chapters in this text lend themselves to prototype models others do not. For example, Chapter 17, Lean Production Models uses prototypes and regional models are ideal, because every production application listed in section 2 of this chapter is based upon a known cavity or mold pattern. C s 13.1 through 13.3 are all examples of production regional models because they contain all the modeling techniques, including regional sections and were produced for use in Chapter17.

13.10 Determining The Mass Of A Model

A regional model has a mass and can be calculated once a material has been assigned. The volume calculation of a region model is moderately accurate because we can find the topology of the model. While the mass (weight) of a solid model is known, we get a mass because we assign a material property to a regional model with the command SOLMAT. If we assign something like aluminum instead of forged iron for C 13.6, the mass will be very different. The mass properties for a regional or solid model, once the topology and material is known, are: Mass, Bounding Box, Centroid, Moment of inertia, Product of inertia, Radii of gyration, Principal Moments and are requested by typing M, B, C, M, P, R or PM.

The SOLMAT command sets the default material from a material in your surface file. If none is listed the command will ask:

Command: SOLMAT
Change/Edit/LIst/LOad/New/Remove/SAve/SEt/<eXit>:N
Enter material name: Mild Steel

You will find other material files in the DLR Associates CAD directory for all of the common manufacturing materials used in Chapter 20.

13.11 Data for Dimensioning Using CP option

In C 13.2 the topology of a model can determine all the necessary data to accurately dimension a lean model. The steps involved in this process are discussed in detail in Chapter 15, for now C 13.2 just illustrates how the various parts of the regional model are presented in working view relationship because dimensions are shown on orthographic views most often. As seen in Chapter 10, orthographic views are automatically generated with filtered XYZ data. More information may be obtained by reference to Chapter 14 (Lean Dimensioning and XYZ Filters).

13.12 Forging Applications

The upper right view port in C13.2 represents a upper and lower forging die and the forged product. The product appears in the upper left view port in a grided measurement matrix. The steps involved in this process are discussed in detail in Chapter 17, for now C13.2 shows how regions can be used to produce a forging die and of course the finished forged part. Regional models are very helpful because often times only an orthographic view of an desired part exists within a directory. This is enough, first pass a section cutting plane through the orthographic view, this will produce one or more sections in the adjacent view. Once this has been done, select this cross-section and use the SOLIDIFY command to produce a region. This region can then be used to produce the required forging model as shown in Chapter 17.

13.13 Casting Applications

In C13.2, lower right view port, the topology of a regional casting model can determine all the necessary data to accurately dimension and tolerance a lean model. The steps involved in this process are discussed in detail in Chapter 16, for now C 13.2 just illustrates how the various parts of the casting model are presented in working view relationship because dimensions and tolerances are shown on orthographic views most often. As seen in Chapter 10, orthographic views are automatically generated with filtered XYZ data. More information may be obtained by reference to Chapter 14 (Lean Dimensioning and XYZ Filters) and Chapter 17.

13.14 Stamping Applications

In C13.3 surface measurement for stamped manufacturing is shown in the upper left view port. A stock guide, shown in the lower left view port guides the material through the punch machine shown in the right view port. See Chapter 17 for more details on stamped surfaces.

13.15 Extrusion Applications

In C 13.3 a typical hot extrusion regional model is shown. A powdered metal is placed in the heated cavity and a punch and die add high pressure to the process. The extrusion molding machine is shown in the right viewport, while a choice of punches and dies are shown in the lower left viewport. In the upper left viewport the surface measurement process for many different extruded shapes is shown.

13.16 Cold Forming Applications

C 13.3 also illustrates two main types of cold forming. They are cold extrusion and impact extrusion called cold heading. In both of these processes a metal slug is shaped in a die and punch as shown in the right viewport. In the upper left viewport the surface measurement of the region is shown. In the lower left viewport sample shapes are shown. For more information on cold forming see Chapter 17, Lean Production Models.

13.17 Deep Draw (Spinning) Applications

C 13.3 shows the deep draw process for repeated impact press work in the right viewport. Deep drawing can also be done on a spinning lathe, the surface measure samples in the upper left viewport illustrate this. A repeated punch press process is shown in the lower left viewport.

The purpose of the applications section of this chapter is to show how the regional modeler is used throughout the industrial production pro- cess (Chapter 17) and in Chapter 20.

13.18 Roll Forming Applications

C 13.3 clearly illustrates how roll forming can produce a cross sectional product. Roll forming, shown in the right viewport, is a continuous, high speed process that shapes strips of material by means of progressive forming roller dies as shown in the upper left viewport. The finished rolled product is shown in the lower left viewport.

13.19 High Velocity Applications

C 13.3, right lower viewport illustrates explosive forming, one of four types of high velocity forming. In explosive forming large panels are lowered into a water vat and placed over the forming die. A holding ring is added and a vacuum assist aids the explosive charge shape the material. The upper left viewport measures the surface shapes after forming, while the lower right viewport shows the preferred finished product. Check Chapter 17 for additional information on this and other uses for regional modeling.

13.21 Chapter Summary

The purpose of this chapter was to concentrate on the fourth major type of lean modeling introduced earlier in this book. A more detailed method of describing objects than wireframe, surface or solids modeling as shown in Chapters 10, 11 and 12, section and profile models use cross sections of solids in 3-D space called regions. The fourth of the four methods is called regional modeling. This was a slightly more difficult form of lean modeling to visualize in the mind's eye. A region model was a cross section of a wireframe, surface or solid corresponding to the shape of the object being modeled. A solid model may consist of a combination of regions, but a region model can not have a combination of solids. A region was unique to its cross sectional shape.

The main types of regional models were:

1. Extruded regions,
    - A. SOLCHP (edits a region primitive)
    - B. SOLCUT (region cutting plane CP)
    - C. SOLEXT (regional ELEV command)
2. Revolved regions,
    - A. SOLREV (rotates a region)
    - B. SOLMOVE (rotates a region or solid)
3. Boolean regions,
    - A. SOLINT (intersections),
    - B. SOLSUB (subtractions),
    - C. SOLUNION (unions), and
    - D. SOLSEP (separations).
4. Shaped regions and
    - A. SOLSECT (cross section)
    - B. SOLFEAT (region from a face or edge)
    - C. SOLPROF (region from a profile view)
    - D. SOLCHP (edits a region primitive)
5. Topology regions.
    - A. SOLMESH (places a polymesh)
    - B. SOLPURGE (removes data)
        1. SOLIN (imports assembly files)
        2. SOLOUT (exports files)
        3. SOLMAT (material assignment)
        4. SOLVAR (sets AME variables)
    - C. SOLUCS (aligns the UCS to the face)

14

XYZ Filtered Dimensioning

XYZ filtered dimensioning is the process of annotating a lean model to show sizes of model features or the distances or angles between faces, edges and features. Features may include holes, slots, bosses, fillets and rounds, chamfers, counter sinks and bores, and many others. Designers and engineers agree that dimensioning is the most time-consuming part of the preparation of a lean model. DLR Associates has simplified the task of dimensioning a lean model as much as possible by automatically calculating distances between selected points and by providing a dialogue box of dimension features for you to begin dimensioning a model. In Chapter 10 we used XYZ filtered database to automatically produce frontal, profile and horizontal (PLAN Command) orthographic working views of a 3-D lean model. Review sections 10.2 and 10.3 if you do not recall this discussion. In this chapter, you will use the same technique to add the dimensions to the orthographic viewports and to the axonometric viewport.

14.1 The Setup Technique

In this section you will prepare the model for dimensioning, you may follow along by downloading the C figures from the CAD directory.

Because most lean models are very complex, as shown in C figure 14.1, we will use a stored model from Chapter 7 as an example. Use the following setup:

1. Select CAD from the main DLR directory.
2. Select the NEW command and create a new drawing based on the bracket.dwg file. In the Create New Drawing Dialogue box, enter:

Create New Drawing
      Prototype. . .   **bracket**
      No Prototype
      Retain as Default

New Drawing Name . .   **bracdim**
                 OK   **Cancel**

bracket in the prototype edit box and bracdim in the new drawing name edit box. The drawing bracket.dwg appears on your screen and can be compared with C figure 14.2.

3. Use command DDLMODES to display Layer Control dialogue box. You can use this box to make the bracket plan layer current. This means that the dimensions for the top view can be turn off or on. C figure 14.2 is an example of the dimensioned layer turned off. C figure 14.3 is an example of the dimensioned layer turned on.

14.2 Defining the Dimension Styles

Dimension variables control the creation and display of dimensions. The dimension styles let you save the current settings of the dimension variables to a named style. The dimension style remains in effect until you change it by:

Command: DDIM

You will need to experiment with this dialogue box. Enter the name default in the dimension style edit box and the current dimension variable settings are saved under the name default. Now enter a new dimension style (nonord). Next select any of the dimension variable boxes to see that dialogue box. If we select Features ... button, the screen should look like the examples shown on the book's CD.

14.3 Linear Dimensions

Linear dimensions are contained within the linear submenu and consist of:1) Horizontal, shown in the lower right of C figure 14.2, 2) Vertical, shown in the upper left of C figure 14.3, 3) Aligned, shown in C figure 14.1, 4) Rotated, shown in C figure 14.1, 5) Baseline, shown in the upper left of C figure 14.4 and 6) Continue, shown in the upper left of C figure 14.5.

You select the object feature to be dimensioned in the horizontal or frontal view. Suppose we wanted to dimension the distance between holes in the horizontal view? Select the center of each hole as shown in the lower right view port of C figure 14.2. This is the feature (hole) location to be dimensioned. The distance is retrieved from XYZ database for bracket.dwg. You will then be asked where to position the information. After you indicate a location, a complete dimension package (extension lines, dimension lines and text) is displayed at that location.

The command prompt area looks like this:

Command: dim
Dim: hor
First extension line origin: cen
Second extension line origin: cen
Dimension line location (Text/Angle): T
Dimension text <32.00>:

But all you did was click LINEAR from the pull-down draw menu shown in C figure 14.1, clicked HORIZONTAL from the linear submenu, and clicked on the display screen center lines. Nothing could be easier to get a 32 millimeter dimension package displayed for you.

14.4 Using Horizontal Dimensions

We need to repeat this point and press process for all the rest of the horizontal dimensions shown in C figure 14.3. In the upper left viewport the 22 millimeter horizontal dimension is placed. In the lower left view port the 20 millimeter horizontal dimension is placed using the same technique of first pulling down the DRAW menu, selecting Dimensions pull-down LINEAR submenu and then clicking the horizontal sub command.

You will note that in C figure 14.3, upper left view port, the centers of the holes are marked by selecting the following sequence: from the draw menu select dimensions, from the dimensions submenu select radial, from the radial select center mark. We can use this command to mark the centers of the holes shown.

14.5 Displaying Vertical Dimensions

To display the vertical dimensions shown in C figure 14.3: select the draw menu, click Dimensions, click linear, click vertical and then select the screen locations for all the vertical dimensions to appear. We used this command to place the 102 dimension in the horizontal view and the 20 in the frontal view of C figure 14.3. Vertical dimensions were used throughout C figure 14.4; in the upper left view port 20, 28 and 14; in the upper right 8 and 16; in the lower left 6 and 12.

14.6 Radial Dimensions

We will add the radial dimensions by pulling down the draw menu, pulling down Dimensions, clicking Radial and then clicking Radius as shown in C figure 14.6. Point at the locations shown in the upper right view port and this will place the dimension package. Even radial surfaces can be dimensioned as shown in the upper left view port of C figure 14.6. In the lower left view port two radii are used to produce a slot and are dimensioned R 10 2 places. In the lower right view port a corner round is dimensioned as R10.

14.7 Dimensioning Diameters

The procedure for showing diameter dimensioning is: DRAW, DIM, RADIAL, and DIAMETER selections. Depending upon the size of the diameter, three options are used as shown in C figure 14.7. In the upper right view port these options are clearly shown; if a diameter is too small it is placed out side the circle with a leader. If the diameter is larger then a dimension is placed inside and the number value placed outside the circle. If the diameter is very large the entire package is placed inside the circle.

In the upper left view port of C figure 14.7 we see surfaces which are diameters dimensioned with horizontal and vertical packages. In the lower left view port diameter locations are dimensioned and finally model features which contain diameters are dimensioned in the correct manner.

14.8 Ordinate Dimensions

The next command item in the dimensions submenu is the ordinate menu selection. This selection has three sub commands: Automatic shown in the upper left view port of C figure 14.8, X-datum which dimensions the X coordinate of feature and Y-datum which dimensions the Y direction of a feature.

Ordinate dimensioning is used in Chapter 17 Lean Production Models and Chapter 20 Manufacturing Models. It is presented here in the order that it appears in the pull down menu items.

14.9 Angular Dimensions

The next command item in the dimensions submenu is the angular menu selection. This selection generates a dimension arc with arrowheads and is shown in C figure 14.9. The placement options are shown in the upper right view port, angular surface dimensioning is shown in the upper left view port, the locations used are shown in the lower left view port and a dimensioned model feature is shown in the lower right view port.

14.10 Preparation of Leaders and Notes

The last item on the dimensions submenu is Leader as shown in C figure 14.10. The placement options are shown in the upper right view port, the placement of notes is shown in the upper left view port, typical locations are shown in the lower left view port and a model feature is dimensioned in the lower right view port.

This chapter introduces XYZ filtered dimensioning commands only, it does not present dimensioning techniques. See Chapter 15 for how to dimension lean models.

14.13 Chapter Summary

XYZ filtered dimensioning was the process of annotating a lean model to show sizes of model features or the distances or angles between faces, edges and features. Features included holes, slots, bosses, fillets and rounds, chamfers, counter sinks as shown in the chapter C figures. Designers and engineers agree that dimensioning is the most time-consuming part of the preparation of a lean model.

DLR Associates has simplified the task of dimensioning a lean model as much as possible by automatically calculating distances between selected points and by providing a dialogue box of dimension features for you to begin dimensioning a model. In Chapter 10 we used XYZ filtered database to automatically produce frontal, profile and horizontal (PLAN Command) orthographic working views of a 3-D lean model. Review sections 10.2 and 10.3 if you did not recall this discussion. In this chapter, you used the same technique to add the dimensions to the orthographic viewports and to the axonometric viewport as shown in C figure 14.2.

15

Dimensioning Lean Models

A lean model is intended to convey size and shape information regarding every detail of the 3-D object displayed. The model may be anything from a ball bearing 1 mm in diameter to a complete robotic manufacturing plant. Without definite specifications expressed by dimensions it is impossible to indicate clearly the design intent that will achieve the successful production of the model. Before the study of each of the sections of this chapter, you will need to down load the CD chapter figures from the DLR Directory.

A correctly dimensioned lean model should represent how the model is to be constructed, manufactured, processed or produced. The dimensions should permit ease of production regardless of the production method chosen. A dimensioned model should deal cautiously with the choice of production. The dimensioned display should be flexible, giving the people responsible for production some latitude as to methods.

Although this sounds simple, it is difficult to practice. The model designer has an intimate knowledge of the details and tries to dimension the model so that production personnel do not have to measure or read from the display to find missing dimensions. The dimensioning practice described in this chapter is that recommended by the Special Interest Group for Graphics (SIGGRAPH), the association for Computing Machinery.

15.1 SIGGRAPH Practices

According to SIGGRAPH, the following list of basic rules or guidelines should be observed in dimensioning any lean model (LM).

1. Each LM dimension must have a tolerance, either applied directly or indicated by a general note. Those specifically identified as reference, basic, or maximum dimensions are exceptions to this rule.

2. Dimensions for size, form, and location of LM features should be complete to the extent that there is full understanding of the characteristics of each feature. Neither scaling (measuring the size of a feature directly) nor assumption of a distance or size is permitted. Undimension diagrams like tool paths are excluded.

3. Dimensions should be shown between points, lines, or surfaces having a necessary and specific relationship to each other or controlling the location of other LM components or mating parts.

4. Dimensions must be selected and arranged to avoid accumulation of tolerances and more than one interpretation.

5. The LM display should define a feature without specifying manufacturing methods (CAM). Thus only the diameter of a hole is given, without an indication as to whether it may be drilled, reamed, punched, or made by any other operation. However, in those instances where CAD, CAM or quality assurance is essential to the definition, it must be specified on the LM display.

6. It is permissible to identify as nonmandatory certain processing dimensions that provide for finish allowance, shrink and other requirements, provided that the final dimensions are given on the LM display.

7. Dimensions should be selected for display to provide required

information. Dimensions must be shown in true profile views and refer to visible outlines.

8. Wires, cables, sheets, rods and other display items that are man-ufactured to gage or code numbers must be specified by linear dimensions (Chapter 14), indicating the diameter or thickness. Gage or code numbers may be shown in parentheses following the dimensions.

9. Surfaces or center lines shown on LM displays at right angles to each other are implied to be 90 degrees apart, without specifying the angle on the display output. According to these practices, C figure 14.1 was a LM. The rest of the C figures in Chapter 14 were not, they were explanatory only. Dimensions on all C figures used in this chapter follow the SIGGRAPH format and the XYZ filtered techniques shown in Chapter 14.

15.2 Display Units and Measurement

Dimensions on all LM displays should be in those units of measurement most compatible with the use of the model. For example:

1.	Scientific	1.55E+01	C figure 15.1
2.	Decimal	15.50	C figure 15.2
3.	Engineering	1' - 3.5"	C figure 15.3
4.	Architectural	1' - 3 ½"	C figure 15.4
5.	Fractional	15 ½	C figure 15.5

These formats can be used with any basic unit of measurement. Decimal is perfect for metric units as will as decimal English units. The common US linear unit displayed on LM is the inch or the millimeter. On most LM displays a note stating:
UNLESS OTHERWISE SPECIFIED, ALL DIMENSIONS INCHES

15.3 Nonlinear Display Notation

Nonlinear is called angular dimensioning and is expressed in degrees, minutes and seconds. These are displayed in the C figures that you downloaded for this chapter.

15.4 Linear Display Notation

In Chapter 14, the XYZ filtering routines were used to check the entire list of LSG (Lean Solid Geometry) commands and to convert the list data to fit a particular display unit size. These linear display units are used to calculate, rotate or manipulate the following items.

1. Dimension. A numerical value expressed in an appropriate display size and indicated on a display area together with lines, symbols and notes to define the linear characteristics.
2. Basic dimension. A numerical value used to describe the theoretically exact size, shape or location of a feature on the LM.
3. True position. The theoretically exact location of a feature established by basic dimensions on the display area.
4. Reference dimension. A dimension usually without variation that is used for information only.
5. Datum. Points, lines, planes or cylinders and other LM shapes assumed to be exact for purposes of computation, from which the location or shape may be established.
6. Feature. A feature is any component portion of a LM that can be used as a basis for a datum.
7. Nominal size. The designation used for the purpose of identification of features or parts of a LM.
8. Actual size. The measured size of the LM.
9. Limits of size. The applicable maximum and minimum sizes allowed by the LM designer.
10. Maximum material condition. The display condition where a feature contains the maximum amount of material within the stated limits of size. For example, minimum hole diameter and maximum shaft diameter are both MMC.
11. Least material condition. Also known as minimum material condition (mmc), the opposite of MMC (largest hole, smallest shaft).
12. Regardless of feature size. Used to indicate a location regardless of the MMC or mmc states of the features.
13. Allowance. The intentional difference between the MMC

limits of mating features (hole and shaft). It is the minimum clearance (positive allowance) or maximum interference (negative allowance) between LM features.

14. Tolerance. The range between MMC and mmc for a single feature.

15. Fit. The display portion used to signify range of tightness or looseness which results from the application of a specific combination of allowances and tolerances in mating LM features.

15.5 Dimension Placement On Lean Models

Dimension packages may be placed in a display area as shown in C figure 15.6. This is a "How to"technique practiced by CAD operators. The "Where" and the "Why" are a different matter. In order to know the where and why you must understand how the LM was designed. A different type of display and dimension placement is required for this as shown in C figure 15.7. Decimal notation is preferred for all dimensions other than reference or nominal sizes. Common nominal size notations are used for such features as bolts, screw threads, keyseats and other standardized notes. When displaying decimal notations, the following why rules should be followed:

1. In plus and minus tolerancing, the specified dimensions contains the same number of decimal places as the tolerance applicable in either plus or minus regions. This rule applies for limit dimensioning also, and tolerances are displayed as shown in the C figures downloaded.

2. Dimensions are specified by means of dimension lines, extension lines and leader lines from a dimension, note or specification directed to the appropriate feature. General notes can convey added information.

3. Numerals indicate the units of measurement. Dimension lines should be displayed parallel to the direction of measurement. The display should meet the COM (computer output to microfilm) requirements.

4. The following should not be displayed as a dimension line: a

centerline, an extension line, a line that is part of the outline of the LM, or a continuation of any of these lines. Exceptions to this appear when automatic ordinate dimensioning is used.

5. The shortest dimension line is displayed nearest the outline of the LM part feature. Where extension lines cross other extension lines, dimension lines or object lines, they are not broken. Cases where extension lines cross arrowheads, a break is permitted.

6. Where a point is located by extension lines only, the extension lines should pass through the point. A special type of extension line is called a leader. It is displayed to direct a dimension, note or symbol to the intended place on the display area. A leader ends in either an arrow (edge placement) or dot (surface placement). Avoid the following display situations: crossing leaders, long leaders, leaders that are horizontal or vertical, leaders parallel to other display items.

15.6 Display of Dimensional Expressions

Dimensional expressions include the directions for the display of dimensions, reference, over all, inside and not to scale. Dimensions on computer display areas are either unidirectional or aligned. Unidirectional are placed to be read from the bottom of the display. Aligned are displayed parallel to their dimension lines. Numerals are read from the bottom or right side of the display area. In both methods, dimensions and notes with leaders are aligned with the bottom of the display area.

The preferred method for indicating reference dimensions is to enclose the dimensions within parentheses. As an alternative, reference dimensions may be displayed using the abbreviation REF placed directly following or under the dimension. Where an overall dimension is displayed, one intermediate dimension is omitted or identified as a reference dimension. Where the intermediate dimensions are more important than the overall dimensions, the overall dimension, if specified, is displayed as a reference dimension.

Dimensions are normally placed outside the outline of the model view. Where directness of application makes it desirable, or where extension lines or leader lines would be excessively long, dimensions are displayed within the outline of the model as shown in C figure 15.7.

15.7 Notes, Tables and Legends

Notes, tables and legends are displayed to furnish certain information that could not be presented in other ways as shown in C figure 15.8. These types of displays are used for such a variety of purposes that it would not be practical to try to establish a procedure for every condition. The following guidelines apply for most notes, tables and legends.

1. All information displayed should be clear, accurate, complete and capable of interpretation inn only one way.
2. Commonly used and understood process terms should be used, try not to use unfamiliar terms or abbreviations.
3. Displays of this type should be simple, brief and concise.
4. Displays should always be read from the bottom of the display area.

15.8 Display of Dimension Limits

A limit note or dimension is used to indicate the largest or smallest dimension allowable as shown in C figure 15.9. The amount of the limit is displayed by a plus and/or minus sign and accompanying legend. The SIGGRAPH practice is to display limits in decimals. See Chapter 18 Lean Dimensioning and Tolerancing for more information on limits.

15.9 Display of Dimension Tolerances

A tolerance is the difference between the limits. The amount of difference is shown in C figure 15.10 for a running and sliding fit. In this example, a component was designed to have a 2.250 diameter shaft having a class RC8 fit to slide in a hole with a nominal diameter of 2.250. For this example the procedure would be:

1. Locate the nominal size range of the hole and shaft in the legend which is 1.97-3.15.
2. Under the column class RC8, the limit range for the hole size runs from .000 to plus .0045, and for the shaft a minus .006 to a minus .009.
3. Since the nominal hole and shaft size is 2.250, the hole may range from 2.250 to 2.2545 and the shaft from 2.244 to 2.241.

The amount tolerance range is displayed from an accompanying legend. The SIGGRAPH practice is to display the tolerance in decimals like a big fraction. See Chapter 18 Lean Dimensioning and Tolerancing for more information on tolerances.

15.10 Display of Dimensioned Sizes and Shapes

The application of sizes to shapes is very apparent when LM display objects must be specified. This is demonstrated in C figure 15.11. The lean profile information and features for LM parts have unique methods of dimensioning.

1. Where the diameters of a number of concentric features are specified as in C figure 15.11, they are sized from an accompanying legend.
2. Lean features in this category include holes, spotfaces, counter sinks and bores, chamfers, fillets and rounds.
3. Slotted holes can be sized in this manner.
4. Key seats are shown in C figure 15.8.
5. All remaining items in this category were shown in Chapter 14.

15.12 Chapter Summary

A lean models in this chapter was intended to convey size and shape information regarding every detail of the 3-D object displayed. The model may be anything from a ball bearing 1 mm in diameter to a complete robotic manufacturing plant. Without definite specifications expressed by dimensions it was impossible to indicate clearly the design intent that will achieve the successful production of the model.

A correctly dimensioned lean model should represent how the model was constructed, manufactured, processed or produced. The dimensions should permit ease of production regardless of the production method chosen. A dimensioned model should deal cautiously with the choice of production. The dimensioned display should be flexible, giving the people responsible for production some latitude as to methods.

Although this sounds simple, it is difficult to practice. The model designer has an intimate knowledge of the details and tries to dimension the model so that production personnel do not have to measure or read from the display to find missing dimensions. The dimensioning practice described in this chapter was that recommended by the Special Interest Group for Graphics (SIGGRAPH), the association for Computing Machinery.

16

Dimensioning & Tolerancing

In recent years there has been a growing tendency to design lean models to solve highly complex computer-aided engineering problems. The Lean Model (LM) is a product of the computer-aided design (CAD) process. The CAD display is intended to be used in computer-aided manufacturing (CAM), and if this process is integrated CADAM (computer-aided design and manufacturing) is the result. With the ever-increasing importance of digital manufacturing processes, the trend in many industries is a gradual shift to an all SI (metric) system of lean dimensioning and tolerancing. The aviation, automotive and electronic industries are now using an all digital system to complete in the world marketplace. The dimensioning system described on this chapter is based upon a LM that is described in millimeters and microns. The micron is used to define the tiny amounts of variation required during manufacturing. The primary advantage that results from using lean dimensioning and tolerancing is the simplification of lean computations.

16.1 Lean Symbols

The lean symbols used in lean model dimensioning and tolerancing are shown in CD Table 16.1. The use of symbols instead of notes provides a number of advantages. CD Table 16.2 will help illustrate that:

1. A symbol has uniform meaning.
2. Symbols are compact, quickly displayed on a CAD output device, and controlled by the LM designer.
3. Symbols are part of the international CAD language, as shown in C figure 16.1 and not subject to a verbal language.
4. Lean lean dimensioning and tolerancing symbols follow the established precedent of other digital systems.

16.2 Angularity

Angularity tolerances are usually the distance in between two parallel planes. A LM designer displays the angularity symbol as shown in C figure 16.2. Angularity is also the condition of a surface or axis at an angle other than 90 degrees that must be part of the LM display. An angularity tolerance indicates one of the following:

1. A tolerance zone is displayed by two parallel planes at the specified basic angle from a datum plane, or axis, within which the LM surface must be presented for viewing.
2. A tolerance zone defined by two parallel planes at the specified basic angle from a datum plane or axis, as shown in the right view port of C figure 16.2. In this example, the feature axis must lie between two parallel planes 0.2 apart which are inclined 60 degrees to a datum plane (A). Angularity applies only to the view on which the symbol is displayed.

16.3 Basic Size

C figure 16.2, left view port shows a part with a surface angular requirement. Note that the symbol is interpreted as:
THIS SURFACE SHALL BE 30 DEGREES- DATUM A

where the 30 degrees is considered a basic size. A basic size must always have a datum plane to reference or place the lean dimension. A dimension placed by a LM designer as BASIC is a theoretical value used to describe the exact size, shape or location of a feature. This requires a tolerance stating the permissible variation from the exact value. The dimension is placed on the LM display and enclosed in a frame or box as shown in C figure 16.3.

16.4 Concentricity and Cylindricity

A concentricity or cylindricity tolerance zone is confined to the annular space between two concentric cylinders as shown in C figure 16.4. A LM designer displays the symbols for concentricity as shown in the right view port of C figure 16.4. The feature axis must lie within a cylindrical zone 0.1 diameter, regardless of feature size.

The cylindricity symbol is placed as shown in the left view port of C figure 16.4. Here the cylindrical surface must lie between two concentric cylinders, one having a radius 0.25 larger than the other. Additionally, the surface must be within the specified limits of size.

16.5 Circularity and Diameter

The diameter symbol was introduced in Chapter 15 Dimensioning LMs and is used to replace the DIA notation found with the decimal inch system. In C figure 16.5 right view port, the diameter symbol is used to indicate two metric diameters of 25 and 14. This means that the first diameter has a maximum allowable distance between axis of datum feature and its axis of .5, while the second has only .1.

Diameters can be held with the circularity symbol shown in the left view port of C figure 16.5. Here the diameter is held within a 0.25 wide tolerance zone.

16.6 Flatness

The flatness tolerance specifies a zone within which the entire surface must lie. A LM designer displays the symbol and notation as shown in C figure 16.6 left view port. Here the surface must lie between two parallel planes 0.25 apart. Additionally, the surface must be within the specified limits of size. Flatness is, of course, related to parallelism as shown in the right view port of C figure 16.6. Here the // symbol is used to define a 0.12 wide tolerance zone for the possible orientation of a feature axis. Flatness is always expressed relative to a datum and in terms of being in parallelism.

16.7 Least and Maximum Material Conditions

The condition of the part at the time of tolerancing may be anywhere within the limit range as illustrated in C figure 16.7. For example, the hole is dimensioned 20 millimeters + 0. and - 0.5. The least material condition (mmc) for that hole is 20. because the hole is the largest allowed and the least material now exists. The maximum material condition (MMC) for the hole is 19.5 because the hole is at minimum size and the most amount of material remains on the LM.

In the case of the boss, the mmc is 30 and the MMC is 31.5. This is because the boss is an external feature and the hole is an internal feature. On all external features the largest dimension allowed is the MMC and the smallest dimension allowed is the mmc. With an internal feature like a hole or a slot, as shown in C figure 16.8, just the opposite is true. Remember, the total weight of the LM determines the MMC or mmc of the model. If the LM is at its lightest weight it is at mmc, if it is at its heaviest weight it is at MMC.

16.8 Parallelism

Parallelism and its relationship to flatness has already been described in Section 16.6. It may also be used for:

1. A tolerance zone defined by two parallel planes perpendicular to a datum plane or axis.
2. The zone between two planes within which the axis of the LM must appear during display as shown in C figure 16.9.
3. A cylindrical zone perpendicular to a datum plane where a display axis must appear.
4. A zone within which an entire surface must appear.

16.9 Perpendicularity

Perpendicularity is shown in C figure 16.10 as the condition of surfaces, axes or lines which are to be displayed at 90 degrees from a datum plane or reference axis. The LM designer displays this symbol to control the LM production operations such as drill, bore, counterbore, countersink or threaded sections as shown in C figure 16.11. Here the thread profile is at MMC, the feature axis must lie within a cylindrical zone 0.3 diameter which is perpendicular to and projects from datum plane A for the 14 specified height.

16.10 Profile of Any Line

The profile of any line symbol looks like an eyebrow as shown in Table 16.1 and demonstrated in the left view port of C figure 16.12. This symbol is used by the LM designer to specify a uniform boundary along the true or basic profile. Profile tolerances can be displayed as:

1. Lean symbols for an appropriate display view of a LM showing the desired basic profile line.
2. All round the profile of a part as shown in C figure 16.13.
3. Segments of a profile that have different tolerances as shown in the top view port of C figure 16.14.
4. Specification profile of a surface between points as shown in C figure 16.14.
5. Specification profile of an outline with sharp corners.
6. Specification of combined profile and parallelism tolerances as shown in C figure 16.13.
7. A profile of a line and size control.

16.11 Profile of Any Surface

The profile of any surface symbol looks like a biscuit as shown in Table 16.1 and demonstrated in the right view port of Figure 16.12. This symbol is used by the LM designer to specify a uniform boundary along the true or basic profile. Profile tolerances can be displayed as:

1. Lean symbols for an appropriate display view of a LM showing the desired basic profile surfaces.
2. All round the surface profile of a part as shown in C figure 16.13.
3. Segments of a surface profile that have different tolerances as shown in the top view port of C figure 16.14.
4. Specification profile of a surface between points as shown in C figure 16.14.
5. Specification of surface profile of an outline with sharp corners.
6. Specification of combined profile and parallelism tolerances as shown in C figure 16.13.
7. A profile of a surface and size control.

16.12 Projected Tolerance Zone

A projected tolerance zone applies to a hole and a mating part, as shown in the lower view port of C figure 16.14. Here the LM feature noted GAGE is a counterbored hole with diameters of 25.4 and 14. A headed fastener is to be assembled into the workpiece labeled GAGE. The fastener at mmc is 24.5 and 13.9. At MMC the fastener is 25 and 14. The fastener is free to float anywhere inside the gage. See Chapter 12 for floating fasteners.

The GAGE is fixed and can not move for ease of assembly. It is important that we label all such conditions with the symbol p so that an interference check can be made as shown in Chapters 10, 11 and 12 of this textbook.

16.13 Regardless of Feature Size

The symbol s is used to describe the condition of floating or fixed fasteners. Floating fasteners work will at any material condition. They may be used anywhere between mmc and MMC (regardless of feature size) and assembly is possible. If one side of the assembly becomes fixed, then some restrictions may apply. For example, the note at true position only may be added. If both sides become fixed, then further restrictions are required and a note at MMC only may be added.

In C figures 16.11 and 16.12, regardless of feature size notes were added in the right view ports of each C figure. This meant that anywhere between mmc and MMC the tolerance applied.

16.14 Roundness

The symbol O was used to describe circularity in C figure 16.5. Circularity is not recognized by SIGGRAPH and the term ROUNDNESS should be used in place of it.

16.15 Runout

The LM symbol for a runout is displayed in C figure 16.5 and is used to define deviation from the desired form of a LM surface detected during full rotation of the LM on a reference display axis. In the left view port of C figure 16.15 circular runout is shown. The display symbol is a single arrow and means that at any measuring point, each circular element of the surface must be within the specified runout tolerance of 0.02 when the LM is rotated 360 degrees.

In the right view port of C figure 16.15 a double arrow is used to indicate that an entire surface must be within the specified runout tolerance zone 0.02 when the LM is rotated 360 degrees.

16.16 Straightness

The straightness symbol is used by a LM designer to indicate a condition where a display element of a surface is a straight line. C figure 16.16 demonstrates two conditions for straightness. In the left view port each longitudinal element (A,B or C) of the LM surface must lie between two parallel lines 0.02 apart where the two lines and the nominal axis of the LM share a common plane. Additionally, the feature must be within the specified limits of size and the boundary of perfect form at MMC (C16.16A).

Waisting, shown in C figure 16.16B or barreling shown in C figure 16.16C, of the surface must be within the straightness tolerance and not exceed the limits of the size feature. In the right view port of C figure 16.16, the derived axis or center line of the actual feature must lie within a cylindrical tolerance zone of 0.04 diameter, regardless of feature size.

16.17 True Position

The true position symbol O is used to describe the exact (basic) location of a point, line or plane. Positional tolerances can be used to show:

1. Location of holes with datum references as figure 16.17.
2. Patterns of holes located by composite tolerancing.   3. Positional tolerancing for coaxial holes as figure 16.18.
4. Tolerancing for symmetry.(figure 16.19 left view port)
5. Spherical feature tolerancing. (fig. 16.19 right view port)

16.18 Chapter Summary

In recent years there has been a growing tendency to design lean models to solve highly complex computer-aided engineering problems. The Lean Model (LM) was a product of the computer-aided design (CAD) process. The CAD display was intended to be used in computer-aided manufacturing (CAM), and if this process was integrated CADAM (computer-aided design and manufacturing) is the result. With the ever-increasing importance of digital manufacturing processes, the trend in many industries was a gradual shift to an all SI (metric) system of lean dimensioning and tolerancing. The aviation, automotive and electronic industries have been using an all digital system to complete in the world marketplace. The dimensioning system described on this chapter was based upon a LM that was described in millimeters and microns. The micron was used to define the tiny amounts of variation required during manufacturing. The primary advantage that results from using lean dimensioning and tolerancing was the simplification of lean computations.

17

Lean Production Models

Lean production modeling with computer-aided process planning (CAPP) software like; AUTOCAM, AUTO CODE, AUTO FAB, AUTO FOLD, AUTO MOLD, AUTO PUNCH and AUTO STAT is a process whereby the engineer describes the production plan for the model geometry, construction, fabrication or the assembly of a LM (LM Model). If the LM is also part of an automated planning process, it is called CAPP. Commonly displayed CAPP lean models always involve parts fabrication. Parts fabrication includes, but is not limited to the following:

1. Casting
2. Forging
3. Extruding
4. Cold forming (extruding and heading)
5. Stamping
6. Deep drawing
7. Spinning
8. Roll forming
9. High velocity forming
10. Machining

In addition to the fabrication processes listed above, CAPP lean models contain items from many of the previous chapters, for example, a CAPP model contains orthographic views arranged in XYZ filtered fashion, section views within regional models, dimensioning from solid models, notation and lettering from chapter 4 and many other concepts presented on lean image processing, storage and lean geometry. These are all used to help describe and document the production process.

17.1 CAD Process Planning for Production

Production CAPP must be understood by the person who prepares the lean production model (LPM). For example, a student who is enrolled in a lean modeling course at Clemson University may also have taken a computer-aided process planning course. In that course the following DLR application packages would have been reviewed: 1) CAD Bill of Materials from C.R. Smolin, Inc.(619-454-3404), 2) ACU.CARV ADS from Olmsted Engineering Co. (616-946-3174), 3) Advanced System 2000 from Compucor (714-779-1795), 4) AMADA 3D Sheetmetal unfolding: US Amada, Ltd (714-739-2111), 5) APPS from Alden Computer Systems (508-744-1314), 6) AUTOCAM from Encode Inc. (603-882-4666), 7) CADmark from Dahlgren Control Systems (415-588-9930), 8) EZMRP material planning from C.R.... Smolin, Inc., 9) FABRICAM from Metalsoft, Inc. (714-549-0112) and 10) TKsolver from Universal Technical Systems, Inc. (815-963-2220)

A typical computer-aided process planning and production model is shown in C figure 17.1. The eight step process planning is shown on the right and the LPM is shown in the upper left viewport of C figure 17.1.

17.2 CAPP for Casting

Casting processes can be classified either by the type of mold or pattern, or by the pressure of force used to fill the mold. Conventional sand, shell and plaster molds utilize a permanent pattern as shown in C figure 17.2, but the mold is used only once. Permanent molds and die castings are made in metal sections and are used for a large number of mold making sessions. Investment casting as shown in C figure 17.1 is a full-mold process that involves both an expendable mold and pattern.

Most castings are produced by filling the mold cavity with a molten metal by the action of gravity. In die casting, the metal is forced into the die cavities under pressure. Some sand, shell and investment molds are filled with molten metal by centrifugal force. The most widely used casting process uses a permanent pattern that shapes the mold cavity when a loose molding material is pressed over the pattern. The compacting of this material is called jolt and squeeze. A typical sand mold, with the various provisions for pouring the molten metal is shown in C figure 17.2. Sand molds consist of two or more sections called a drag, cope and cheek (bottom, top and intermediate) sections. The molten metal is poured into the sprue (opening in the sand), and connecting runners called gates provide flow channels for the metal flow to enter the mold cavity. Risers are located over the heavier sections of the casting as shown in C figure 17.2. These risers fill with molten metal and provide reservoirs to compensate for the contraction of the metal during solidification. The sections are held together with alignment pins and holes or with clamps. The core is held in place with metal pins called chaplets and core prints. Chaplets fuse into and become part of the casting. The core is removed after cooling and forms an opening in the cast part. Once the cast part is solid, it is removed from the sand mold, the core is removed along with the sprue, gates, runners and risers. A sand type mold usually will require more than one type of sand. For example:

Green sand is a mixture which contains 2 to 8 % water is used directly over the molding board. It forms a bounding to support the rest of the mold cavity.

Dry sand is the sand directly in contact with the pattern which forms the mold cavity. Usually the pattern is split along a centerline and mounted on a metal plate. This plate is called the match plate. A match plate extends beyond the flask, over the alignment pins and is heated to remove some of the water contained in the green sand. The dry sand does not fall apart because dry sand mixtures also contain various organic binders, together with clay.

CO2 sand is used for cores. Cores are often molded with special sand containing 2 to 5% liquid sodium silicate. This binder becomes very hard when gassed with carbon dioxide. This process is robot computer automated and is recommended to be used in place of another core technique which requires baking in an oven. An oven baked core production process can be automated - but these cores are thrown away after use, while the CO2 sand is mulled (mixed) and used again.

Shell sand is refactory sand used in shell molding and is bonded by a thermosetting resin that forms a relatively thin shell over the pattern. Here again a match plate is used to heat the sand. Green sand is then used to back-up the shell covering.

Of course other substances besides sand are used to form CAPP casting molds, some but not all types are listed below:

1. Plaster, consisting of gypsum with various amounts fillers.
2. Investment, (wax, plastic, frozen mercury) as figure 17.1.
3. Die, where molten metal is forced into permanent molds.
4. Centrifugal, where the mold is rotated rapidly.

17.3 CAPP for Forging

Forging consists of deforming a heated metal ingot, bar or billet by squeezing or hammering. Practically all ductile metals can be forged if heated to a plastic state. In order to better understand CAPP displays, the types of forgings are:

Closed die. Forgings that are made by hammering or pressing metal until it conforms closely to the shape of the die.

Open die. Forgings are made by hand with simple tools like hammers and flat dies. The metal is progressively worked to shape by hand and depends upon the skill of the forger.

Blocker type. Forgings that are a combination of closed and open dies are called blocker. They require closed dies but are worked by hand.

Close tolerance. Also known as draftless, precision forging; this type of forging is used for fine quality tool parts.

Upset. These types of forgings have a certain amount of material pushed back on itself. This may occur as a flange at one end or as collars or bosses along the body.

CAPP forging displays are used to describe the type of production that is to be used. For example, C figure 17.3 upper viewport illustrates a parting line (sometimes called a flash line). This is where the forge dies meet and separate. The parting line, or lines, need not be in the same plane as shown in the simple locked parting line in C figure 17.4. Here a parting plane (also called forging plane or die plane) is shown. It is a plane perpendicular to the direction of pressure. This may or may not be in the same plane as the parting line. An upset forging, for example, has two parting planes. The lower viewport of C figure 17.3 shows die closure. This is the variation that occurs when forging dies do not close completely.

17.4 CAPP for Extruding

Extruding is a special form of forging where hot metal is forced through a die to form a shape having a cross-section matching the opening in the die as shown in C figure 17.4. Extrusion presses are usually horizontal, hydraulically operated. This type of press consists of a heated container or holder for a billet, a ram that applies pressure to the billet, a hardened die and a shear to cut the material to length. A die can be visualized by looking at C figure 17.4 again and saying to yourself, this is the material being pushed through the face of the die opening. Most dies are round blanks the diameter of the billet (raw material) and are inserted as shown in the upper left viewport.

Temperature, pressure and extrusion speed depend on the metal and cross-section shape of the die. The most popular metal is aluminum, its heat range is 400 to 1600 degrees F; pressure 30,000 to 160,000 PSI; speed 1 to 1000 fpm.

17.5 Cold Forming

Two main types of cold forming can be used. They are extrusion and cold heading. These processes are also called impact extruding. Cold extruding is a process in which a metal slug is shaped in a die and punch as shown in C figure 17.5. Most parts formed by the process are characterized by fairly long tube-like cross sections as displayed in C figure 17.5. Cold heading was developed primarily to make nuts, bolts, screws and other type fasteners illustrated in Chapter 19.

In cold heading, a blank is fed into a die and worked by a series of die motions so that it assumes the shape of varying diameters. The process generates almost no scrap and is therefore economical once tooling costs have been recovered.

17.6 Stamping Processes

Stamping of material includes cutting, bending or forming done at room temperatures. Many types of presses are used for performing stamping, ranging from small bench presses to large automatic computer controlled presses. In order to create CAPP stamping displays, the user must understand and be able to describe:

Shearing. The production of cut sheets form industrial presses containing knife punches.

Cut off. Cutting strips to length in a press.

Blanking. Shown in figure 17.6 produces accurate punched cut portions.

Notching. Separate operations for shaping the edges of punched parts as shown in figure 17.6.

Lancing. Punching which leaves attachments to the parent material (tabs) shown in the upper left viewport of figure 17.6.

Hemming. Bending material back on itself to form a smooth edge.

Flanging. Forming the edge of material at the 90 degree bend.

17.7 Deep Drawing Process Planning

CAPP deep drawing and stamping are production processes often displayed interchangeably, yet the two processes are basically different. In deep drawing, material is held between a pair of rings as shown in figure 17.7 and then made to flow over a punch of the desired shape. Figure 17.7 indicates that parts can be formed to greater depths than a stamped part because movement of the material is controlled. The drawing operation can be done either by a pressing or a pulling action. Pulling actions require open dies.

The pot and pan industry depends upon deep drawing presses and regional modeling as shown in Chapter 13. Figure 13.13 indicates the great variety of pans that can be formed with this production process.

17.8 CAPP Spinning Displays

Spinning is one of the most simple CAPP production methods to control. The entire process is done on a CNC lathe. It is the process where a flat disk of material (aluminum, copper, magnesium, brass or stainless steel) called a blank is shaped over a mandrel as shown in figure 17.8. The mandrel is located on the lathe rotating spindle held by tailstock pressure. The material is placed between the mandrel and tail stock pressure plate and spun onto the face of the mandrel. The mandrel can be thought of as a mold which catches the spinning material. The material matches the shape of the mandrel surface. Several mandrels are shown in the lower left viewport of figure 17.8.

17.9 Roll Forming Process

Roll forming is a continuous, high production process that shapes strips of material by means of progressive forming rolls as demonstrated in the upper left viewport of C figure 17.9. In this CAPP display a series of rollers produces shapes of uniform cross-section with close tolerances. If each of the stations shown could be displayed separately as a regional model then C figure 13.14 of Chapter 13 would be the result. It is more typical to view a CAPP display as shown in figure 17.9, however.

17.10 High Velocity Forming

CAPP high velocity forming shown in figure 17.10 is used to form large industrial parts. Originally called HERF (high energy rate forming) because of the large explosives used in the process. This process is some times awkward because it is the high deformation velocity - not energy, which does the forming. Also included in HERF is:

Explosive forming. Shown in the right viewport figure 17.10.

Electrohydraulic forming. A combination of a pressure cylinder, a plunger, a port, a goose neck feed into a closed die set off by a high energy source, all done under water.

Pneumatic mechanical forging. CAPP displays used extensively in aerospace applications, but other industries have been using this.

17.11 CAPP Machining Display

CAPP displays for forming parts by metal removal is such a large endeavor that an entire chapter can be devoted to this. In this chapter C figure 17.11 deals only with a drill press, for other methods see Chapter 20. In Chapter 20 lean models are used for forming of parts by metal removal that produce shapes with smooth surfaces and dimensional accuracies not generally attainable by the methods used in this chapter.

For this reason, most engineers and technicians will sometimes limit themselves to only this production technique. The truth is that most lean models are a combination of one or more methods in this chapter. For example, castings and forgings are machined to produce the final product. Computer-aided machining introduced in this chapter and covered in detail in Chapter 20, is best used for low and medium production runs. For extremely large production runs CAM is not cost affective. For large-volume production part runs, a machine like a screw machine is preferred. Lathes, drill presses, milling machines and grinders are also used to perform medium run CAM productions.

Screw machines can do turning, threading, forming, facing, boring, cut off, burnishing and specialized milling, slotting and cross drilling. Like a lathe, all machining is done with the axis of the part. Examples of machined parts are found throughout Chapter 19, refer to them now.

17.12 Chapter Summary

Lean production modeling with CAPP software (AUTOCAM, AUTOCODE, AUTOFAB, AUTOFOLD, AUTOMOLD, AUTOPUNCH and AUTOSTAT) was used throughout the chapter and in the review problems just completed. This was a process whereby the engineer or technician describes the production plan for the model geometry, construction, fabrication or the assembly of a Lean Model. If the lean model is also part of an automated planning process, it was called Computer-aided Process Planning (CAPP). Commonly displayed CAPP lean models always involved parts fabrication. Parts fabrication included, but is not limited to the following:

1. Casting
2. Forging
3. Extruding
4. Cold forming (extruding and heading)
5. Stamping
6. Deep drawing
7. Spinning
8. Roll forming
9. High velocity forming
10. Machining

In addition to the fabrication processes listed above, CAPP lean models contained items from many of the previous chapters, for example, a CAPP model may contain orthographic views arranged in XYZ filtered fashion, section views within regional models, dimensioning from solid models, notation and lettering from chapter 4 and many other concepts presented on lean image processing, storage and constructive geometry. These were all used to help describe and document the production process.

18

Lean Gear and Cam Models

The lean design of gears and cams for use as prototype models is often taken for granted because of their apparent simplicity. As shown in the dimensioning chapters, there exists a design file within the software used for this textbook. It is loaded and used to display simple gear and cam 3-D models. This is useful to place gears or cams on shafts, one shaft turns (driver) and causes the other to turn by means of two bodies in pure rolling contact. Cams are often designed on the basis of this principle. If the speed ratio of the lean model must be exact or a rotary motion must be transferred as a rotation instead of a linear motion, toothed wheels called gears are used in place of a cam. The shafts of gears do not have to be parallel, and are often perpendicular. Special types of gears are displayed for shafts that are not parallel as shown in C figure 18.1. Here the miter or spiral solid model is used for perpendicular shafts and the worm gear (involute helicoid thread form) makes this nonparallel shaft location possible.

A cam can be displayed as a plate, cylinder or other solid model with a surface of contact so designed as to translate rotary motion to linear motion as shown in Chapter 11, C figure 10. The cam is mounted to the driving shaft, which rotates about a fixed axis (see chapter 9). By the cam rotation, a follower is moved in a definite path. This is one of the applications for this chapter introduced in Chapter 9. The follower may be a point, roller or flat surface. The follower is attached to another part of the lean model as displayed in Chapter 20.

18.1 Types of Gear Displays

C figure 18.1 illustrates four common types of solid model displays. In Chapter 9, special attention was paid to the fact that rolling bodies may be used to connect axes that are parallel, intersecting or neither. The same situations arise in the use of gears, and special names are given to the display of gears according to the Lean model (LM) for which they are used.

18.2 Gear Terminology

The terminology used in this chapter and listed in C Table 18.1 will be easier to understand if C figure 18.2 is referred to as needed. When gears mate to transfer motion additional terms are useful, these are listed and demonstrated in C figure 18.3.

18.3 Display Relationships

LM designers prefer that all gear teeth be displayed with a slightly curved surface. This is known as an involute tooth form. This display form is necessary to assure simple clearance as the teeth mesh as shown in C figure 18.3. If the teeth had straight sides, they would become damaged within a short time if a large amount of clearance was not provided. This display form is common to all types of gears, with the exception of worm gears.

Another common display fundamental is that all gear teeth are spaced by dimetral pitch (DP). In a display, DP is the storage location for the ratio of the number of teeth to the number of display units in the pitch diameter of the gear and would be found in C Table 18.2 as:

DP = N/PD
where N is the number of teeth and PD the pitch diameter.

There are three formulas for finding DP as shown in Table 18.2. You may also find pitch diameter (PD), outside diameter (OD), number of teeth (N), addendum (A), working clearance (WK), chordal addendum (CA) or chordal thickness (Th).

18.4 Gear Elements

C figure 18.3 shows a partial pair of external spur gears in mesh with each other. Since these are the simplest form of gears, the following discussion will be based on this type of gear. Spur gears were the first gear type (SOLGEAR) provided by programmers. In this software approach to providing a library of solid gear models, it was the first logical step. Although there are currently many other types of gears used in LM designs, spur gears are the most common. It must be kept in mind, however, that the basic gear elements of this book are designed to apply to the four common types of gears shown in C figure 18.1.

The advantage of four types of gear model elements include ease of manufacture, reliability, minimum redesign and ease of design assembly. Because of their design simplicity and library availability, the four gear types can be specified in may types of metallic and nonmetallic materials. Their parallel tooth design enables the gear blanks to be made and displayed in large diameters and broken down or split for assembly as segments in a production application (see chapter 17).

18.5 Computation of Speed Ratios

The profiles of spur gear teeth must be such that the speed ratio is constant. A wireframe prototype display as shown in Chapter 10 section 6 is generated for each tooth size displayed. The various sizes are then simple to display. But because of their simplicity, they have some disadvantages. Because of the minimum number of teeth in contact between the pinion and driven gears, they are usually designed only for slow or medium-speed applications. As shown in C figure 18.4, spur gears are displayed as regular gears and also straight gears called racks. The racks are normally designed to transmit rotary motion from the pinion into linear motion.

Now that the wireframe profile for each of the common DP is stored, the LM designer may display working spur gears. Such a pair is shown in C figure 18.5. Here the larger gear has 60 teeth and the smaller 30. The 60 tooth gear will turn with the 30 tooth gear. As the gear blank turns through (360/60) degree segments, another wireframe tooth is displayed from memory. In this manner, each tooth of the larger gear is displayed as it rotates in mesh. If the center locations of the mounting shafts are properly placed, the teeth of the smaller gear blank may be generated in the same manner. As one gear blank is pushed by another tooth through (360/30) degree segments, the display uses the DP for that gear.

As the display is being generated, a viewer will note that the teeth on one gear blank will seem to push the teeth on the second blank until the gears have turned so far around that two mating teeth swing out of reach of each other or come out of contact. But before these two teeth come out of contact, another pair come into contact so that one gear will continue to drive the second gear. For the pinion to make a complete revolution, each of its 30 teeth must be pushed past the centerline. Therefore in order to display both gears, the pinion must be turned twice.

18.6 Display of Gear Elements

Four types of prototype wireframe gear tooth shapes are needed for this chapter:

1. Spur - shown in C figures 18.3 and 4.
2. Bevel - shown in C figure 18.5.
3. Spiral - shown in C figure 18.6.
4. Worm - shown in C figure 18.7.

18.7 Display of Gear Teeth

When the wireframes have been stored and the circular pitch, backlash, addendum and clearance have been added to the prototype displays they look like C figures 18.3 through 18.8. By viewing the gear teeth, the LM designer checks for such things as tooth curves that have flanks that extend into the root line. This will result in a weak tooth, and the designer avoids this by placing a small fillet at this junction. The size of the fillet is limited by the arc of a circle connecting the root line with the flank and lying outside the actual path of the end of the face of the mating tooth. This path of the end of the face is called the true clearing curve. This curve is the path traced by the outermost corner of one tooth on the plane of the other gear.

18.8 Involute Gears

The form of the gear display most commonly given to computer analysis is that known as the involute of a circle. Gear teeth constructed with this curve will conform to those shown in C figure 18.5. This C figure consists of all the parts used in a typical LM. Notice that a choice of three different gear blanks are shown in the lower left view port. Involute spur gears can be arranged in a number of different patterns. C figure 18.3 is a display of a pinion and rack in mesh. No new principles are involved since the rack is merely a spur gear whose radius of the pitch circle has become linear. The base line of the rack is tangent to the pres-sure angle. For this reason the sides of the teeth of the rack will be straight lines perpendicular to the line formed by the pressure angle. In C figure 18.4 the addendum on the rack is displayed in mesh as much as the pinion will allow.

18.9 Cycloidal Gears

Cycloidal gear teeth as shown in C figure 18.6 are designed by a software system in which the faces of the teeth are computed as epicycloids gene-rated on the pitch circles. The flanks of the teeth are then hypocyloids generated inside the pitch circles. Few, if any, spur gears are designed with this analysis in mind. It is best used in bevel and certain types of spiral gears. The best known use is in the case of high-tolerance gearing, where interchangeability can be a problem. For these cases, the cycloidal system is ideal because the same describing circle can and must be used in generating all the faces and flanks. The size of the describing circle depends on the properties of the hypocycloid curve which forms the flanks of the teeth. If the diameter of the describing circle is displayed to be half of the pitch circle, the flanks will be radial, which displays a com- operatively weak tooth at the root. If the describing circle is entered smaller, the hypocycloid curves away from the radius and will display a strong form of the tooth. However, if the describing circle is input too large, the hypocycloid will curve the other way, passing inside the radial lines as shown in C figure 18.7.

From the display of C figures 18.6 and 18.7, two versions of cycloidal systems have been used. A practical conclusion would appear to be that the diameter of the describing circle should not be more than half that of the pitch circle for bevel gears and more than one half for spiral gears.

18.10 Hypoid Gears

Hypoid gears are shown in C figure 18.8. This prototype gear is a special design resulting from an effort to obtain satisfactory gears for connecting nonparallel and nonintersecting shafts. The appearance of the teeth shown in C figure 18.8 are generated by the use of rotary translation (epicycloids and helix). The pinion of a pair of hypoid gears is often called the worm and its teeth resemble screw threads. The driven gear is larger than the driven gear in cycloidal gears with the same number of teeth. In this respect worm gears are stronger than spiral gears. Hypoid gears, because of the helix translation, have a continuous pitch line contact and a larger number of teeth in contact than cycloidal. Hypoid gears are quieter and wear longer than cycloidal gears.

From the three major gear teeth; involute, cycloidal and hypoid - several applications can be designed and displayed:

1. External spur - C figure 18.5
2. Rack and pinion - C figure 18.4
3. Annular and pinion
4. Mated interchangeable - C figure 18.3
5. Stepped
6. Twisted spur or helical
7. Herringbone
8. Pin

9.   Bevel (crown, internal, spiral)
10. Skew - C figure 18.8

18.11 Types of Cam Models

A cam and its follower form an application of the principle of transmitting motion by direct sliding contact, as presented in Chapter 9. As in the case of gears, various situations arise for the use of cams, and special names are given to the display of cams according to the situation for which they are designed. Gears were based upon wireframe modeling shown in Chapter 10. Cams are based upon regional modeling shown in Chapter 13.

C figure 18.9 is a typical plate cam displayed as a regional model. C figure 18.10 is a typical cyclindrical cam display. Many LM designers depend largely upon cams to five motion to the various parts of the completed lean model. nearly all cams are designed for the special purpose intended. This being the case, speed ratios are not the desired output: rather, a cam assigns a certain series of definite positions that the follower is to assume while the driver occupies the corresponding series of positions.

The relationship between the successive positions of the driver and follower in a cam motion are displayed as a displacement diagram and are placed directly below the cam profile in both C figures 18.9 and 18.10. The abscissae of these two diagrams are linear distances arbitrarily chosen to represent angular motion of the cam, and the ordinates are the corresponding displacements of the follower from its initial position.

18.12 Display of Cam Profiles

A cam imparts motion to a follower defined by the displacement diagram so that its movement is constrained to move in a plane that is perpendicular to the axis about which the cam rotates. This type of cam is best defined as a flat plate shown in C figure 18.9. Cams may also occupy a plane coincident with or parallel to a plane in which the cam rotates. This type of cam is best displayed as a cyclindrical as shown in C figure 18.11.

18.13 Follower Motions

A LM design often requires that a cam transmit a definite displacement in a definite period of time. These movements are called rise, dwell and fall and are shown. In addition the rise or fall can be called uniform velocity, harmonic motion or uniform acceleration.

18.14 Chapter Summary

The lean design of gears and cams for use as part of a LM can often be taken for granted because of their apparent simplicity. As shown in the dimensioning chapters, there existed a design file within the software used for this textbook. It was loaded and used to display simple gear and cam 3-D models. This was useful to place gears or cams on shafts, one shaft turns (driver) and causes the other to turn by means of two bodies in pure rolling contact. Cams were often designed on the basis of this principle. If the speed ratio of the lean model must be exact or a rotary motion must be transferred as a rotation instead of a linear motion, toothed wheels called gears were used in place of a cam. The shafts of gears do not have to be parallel, and are often perpendicular. Special types of gears were displayed for shafts that were not parallel. Here the miter or spiral solid model was used for perpendicular shafts and the worm gear (involute helicoid thread form) made this nonparallel shaft location possible.

A cam can be displayed as a plate, cylinder or other solid model with a surface of contact so designed as to translate rotary motion to linear motion as shown in Chapter 11, C figure 10. The cam is mounted to the driving shaft, which rotates about a fixed axis (see chapter 9). By the cam rotation, a follower is moved in a definite path. This was one of the applications for this chapter introduced in Chapter 9. The follower may be a point, roller or flat surface. The follower was attached to another part of the lean model as displayed in Chapter 20.

19

Lean Fastener Models

A lean model uses fasteners to show the size and shape information regarding every detail of the 3-D object to be assembled. The completed model is usually an assembly display showing what type of fasteners are supplied from an approved supplier and which are unique to the lean model (LM) and must be produced or manufactured locally. Approved suppliers are members of the Industrial Fasteners Institute (IFI).

Since 1999 a LM designer can select IFI fasteners directly from the web site www.industrial-fasteners.org . In June of 1999 the Fastener Quality Act (FQA) was signed into law. This act greatly reduced the cost of world wide threads and fastener interchangeability used by American industry in its operations.

This chapter along with its illustrations is presented as an aid to proper design and display of fasteners or fastener details that can not be down loaded from the IFI. This keeps the emphasis on fastener details for LM design and not fastener displays. The chapter is therefore about: 1) Permanent fasteners (welding, rivets, impact screws, etc.), 2) Assembly fasteners (keys, pins, stud bolts, etc.) and 3) Application of specially designed fasteners in LM displays.

19.1 IFI Members And Practices

Membership in IFI is by division as shown in C T19.1:

1. Industrial Products
2. Aerospace Fasteners
3. Automotive Industrial Fastener Group
4. Suppliers Advisory Council

According to IFI, the following list of basic guidelines should be observed in designing any fastener for a lean model.

1. Each LM fastener must have a tolerance, either applied directly or indicated by a general note. Those specifically identified as IFI vendor supplied are exceptions to this rule.
2. SI dimensions for size, form, and location of fasteners should be complete to the extent that there is full understanding of the character-istics of each fastener. Neither scaling (measuring the size of a fastener directly) nor assumption of a distance or size is permitted. Assembly diagrams are excluded.
3. Fastener locations should be shown between points, lines, or surfaces having a necessary and specific relationship to each other or controlling the location of other LM components or mating parts.
4. Fasteners must be selected and arranged to avoid accumulation of tolerances and more than one interpretation.
5. The LM display should define a fastener without specifying manu- facturing methods (CAM). Thus only the diameter of a threaded hole is given, without an indication as to whether it may be tapped, turned on a lathe, self-tapped, or made by any other operation. However, in those instances where CAD, CAM or quality assurance is essential to the definition, it must be specified on the LM display.
6. It is permissible to identify certain processing fasteners such as spot welding, dip soldering or other joining that provide for finish allowance, shrink and other requirements, provided

that the final fastener notations are given on the LM display.

7. Fasteners should be selected for display to provide required information. Fasteners must be shown in true profile views and refer to visible outlines.

According to IFI practices, C figure 19.1 illustrates a LM display. Fasteners on all C figures used in this chapter follow the IFI format and are the courtesy of the Industrial Fasteners Institute.

19.2 The Fastener Industry

Today, it is estimated that the United States fastener industry operates 380 major manufacturing facilities with 44,000 employees and a total US sales of 7.5 billion dollars. As currently constituted, US fastener production is strongly tied to the production of automobiles, aircraft, appliances, construction and agricultural machinery and equipment, and the construction of commercial buildings and infrastructure. More than 200 billion fasteners are consumed annually in the US - 26 billion by the auto industry alone. C Table 19.1 data from the US Department of Commerce more closely describes the current industrial fastener market.

19.3 Fastener Terminology

Fastener terminology is best learned by a series of illustrations. These illustrations are used to show the various types of fasteners available from the IFI and how to display those that you will design as part of a LM. Commonly used fasteners are shown in C figure 19.1, these and others will now be presented for further study and illustration.

BOLT. Shown throughout C figure 19.1, and in close up detail in C figure 19.2, this is an externally threaded fastener designed to go through holes in assembled parts of a LM, and normally tightened or released by torquing a nut onto lock washers. Additional information is available by logging on the IFI web site.

SCREW. Shown in C figures 19.1, upper left view port is a flanged 12-point screw. As shown in C figure 19.3 this is also an externally threaded fastener capable of being inserted into holes in assembled LM parts or mating with a preformed internal thread. Other types of screws are: Hex head cap, hex socket cap, or those forming thread own thread, and of being tightened or released by the torquing of the head.

STUD. Shown in C figure 19.4, upper left view port, is a cylindrical rod of moderate length, threaded on either ends, usually it replaces the bolt and screw shown earlier. The stud requires a nut, two nuts, or a nut and threaded section to mate or it may be welded onto a surface. It is therefore, more widely used in LM designs than either the bolt or screw.

RIVET. Shown in C figure 19.4, this is a headed fastener of malleable material used to join parts by inserting the shank of the rivet through aligned holes in each piece and formed into a head by upsetting.

TAPPING SCREW. Shown in C figure 19.4 and in C Table 19.2, this is a hardened screw designed to cut its own thread in a plain drilled hole. Tapping screws are used on sheet metal, thin materials of all types, and in other permanent joining applications.

DRIVE SCREW. Similar to the tapping screw, except it is designed with an extremely fast lead thread which permits it to be put into place without the necessity of drilling.

LOCKING SCREW. Shown in Table 19.3, is an externally threaded fastener which has a special means within itself for gripping an internal thread so that rotation of the threaded parts, relative to each other is resisted as in the eight steps of C Table 19.3.

19.4 Fastener Parts

A fastener used in a LM is a mechanical device designed specifically to hold, join or otherwise fasten together two or more parts of the LM. The fasteners described thus far have been for assembly purposes only, many others will be shown as we progress. For now, we must stop a moment and consider the parts of a fastener that must be included in a lean display. They are:

FASTENER TYPE. The type of fastener must be noted on the display. The various types are usually cataloged in the directory and recalled when needed. A list of types that should be included in the CAD directory are:

1. Insert, an internally threaded bushing or thread reinforcement.
2. External torque, shown in C figure 19.2.
3. Internal torque, which is assembled using an internal wrench in a socket or recess in the fastener shown in the lower right view port of C figure 19.4.
4. Aircraft, toleranced fastener used in high vibration applications.
5. Semifinished, lower toleranced than an aircraft fastener with only the threads finished to fit closely.

NOMINAL SIZE. The LM designation used for the purpose of general SI identification. For threaded sections it is usually the metric major diameter. For unthreaded fasteners it is usually the basic metric body diameter.

LENGTH. The length of a LM fastener is always measured parallel to the display axis. The length of a headless (stud) fastener is the distance from one end to another. The length of a headed fastener is from the base of the head to the end of the fastener.

SHANK. The portion of a LM where the headed fastener lies between the base of the head and the end of the fastener. Or it is the part of the fastener which will contain the threads, head location and body.

BODY. The portion of a LM that does not contain threads. Bodies of fasteners may be reduced diameter, externally relieved, internally relieved or contain shoulders.

SHOULDER. This is the enlarged portion of the body, usually next to the head. A shoulder can be a bearing face for a washer.

WASHER FACE. This is displayed as a circular boss on the bearing surface of a bolt or nut.

DRIVING RECESS. Shown in C figure 19.4 lower right view port.

GRIP. The thickness of material or parts which the fastener is designed to secure when assembled should be labeled grip.

19.5 Displaying Threaded Fastener Parts

The display of threaded parts on a lean model is represented by a ridge of uniform section (regional model) in the form of a helix on the external or internal surface of a cylinder as shown in C figure 19.5. However, threaded sections also involve a helix in the form of a conical spiral as illustrated in Table 19.4.

Many excellent fastener designs can be down loaded directly from the IFI-500 standard and can be displayed as a part of a lean model with little effort on the part of the designer. Only the notes and dimensions are changed from one thread display to the next. The common types of displays are:

1. Single thread fasteners, shown in C figure 19.6 have a lead equal to the pitch.
2. Multiple thread fasteners, shown in C figure 19.7.

19.6 Display Form of a Thread

The theoretical profile of a thread is shown in C figure 19.8 and contains:

1. Crest, is the outermost tip of the thread.
2. Root, is the bottom of the thread.
3. Flank, connects the crest with the root.
4. Pitch, is measured parallel to the thread axis between correspon-ding points of the thread form and appears in a thread note (M30 X 1.5-5g) as 1.5.
5. Lead, is the distance that a thread moves axially in one compete revolution.
6. Lead angle, is the helix of the thread at the pitch line.
7. Thread angle, is the lead angle plus the helix angle.
8. Major diameter, is the largest diameter of the thread and appears in the thread note (M30 X 1.5 - 5g) as 30.
9. Minor diameter, is the smallest diameter of the thread.
10. Pitch diameter, is the imaginary cylinder whose surface passes through the thread profile in such a way as to make the widths

of the thread equal to the depth. This is always toleranced and appears in the thread note (M30 X 1.5 - 5g) as 5.

11. Allowance, appears in the thread note (M30 X 1.5 - 5g) as g. This is the interference allowed between mating threads and can be: e = large allowance for external threads, E is internal; g = small allowance for external threads, G is internal; h = no allowance for external threads, H is internal.

The allowance is placed after the tolerances such as: 4 = low tolerance, 6 = normal and 8 = high tolerance.

19.7 Specification of Fastener Material

Because modern industrial fasteners are available in almost any material, the specification for lean notation is practically unlimited. The key to material specification for fasteners is in knowing where the fastener will be used. Usually the use can be determined by answering the following questions:

1. Will the fastener be subjected to corrosives?
2. Will it experience high or low temperature ranges?
3. Is weight critical?
4. Should it be nonmagnetic?
5. How much should it cost?
6. Is it permanent, assembly or temporary life span?

The problem of corrosion can be met by using a protective coating or finish. If a fastener will experience a wide range of temperatures, a LM designer can meet the specifications by cold working or heat treatment. A weight reduction is possible with nonferrous metals. It also can be nonmagnetic in this manner or the fastener may be made form nonmetallic materials. in an elecric motor, a fastener or a magnetic material would not be used next to the coil windings.

When specifying a fastener, always be as general as possible, never over specity chemical properties. These types of specifications add to the cost and availablity of the fastener. The last item to consider is the life span. If the fastener is temporary (staple in a carton) it is thrown away. if the same use is designed for assembly, an adhesive could be used to fasten the carton. If a permanent carton fastener is to be designed, then quick acting fasteners like snaps or pins can be used over and over again.

19.8 Common Fastener Materials

The greatest number of fasteners are made of steel (SAE 1010). Machine screws, carriage bolts and other similar fasteners might use SAE 1018, 1020 or 1021. High strength bolts, studs, nuts and cap screws use SAE 1038.

Aluminum is ideal for applications where high strength-to-weight ratio and corrosion resistance is required. Cold formed bolts, screws, rivets and nuts use 2024-T3 alloys. Brass, copper, nickel and titanium are the other common fastener materials. A titanium fastener has great strength but should not be used next to magnesium because of galvanic corrosion. The choice of a fastener material is often dependent on the material to be joined.

19.8 Set Screws

Set screws are displayed on a lean model as semi-permanent fasteners to hold a collar, sheave or mechanical element on a shaft. In comparison with other fasteners, a set screw is essentially a compression device selected from a table listing such as shown in C Tables 19.5 or 6. The forces acting on the shaft produce a strong clamping action that resists motion between assembled parts. The only problem with using a set screw in a lean model is to select the best combinations of screw based on form, size and point style. The holding power can be determined from C Table 19.5. The various types of points can be selected from C Table 19.6.

19.9 Pins

Pins used as fasteners offer an inexpensive and effective approach to assembly where loading is primarily in shear. Pins are either semi-permanent, like et screws or quick release. Semi-permanent pins are called machine pins and are selected from C Table 19.7 as:

1. Hardened and ground dowel pins.
2. Taper pin.
3. Clevis pin.
4. Cotter pin.

19.10 Spring Clips and Rings

Spring fasteners are particularly important to the mass producing industries. Clips and rings perform multiple functions, often eliminate the handling of several smaller parts and therefore, reduce the assembly costs. C figure 19.9 is an example of the dart spring clip used in four different configurations. C figure 19.10 is the cable and tube clip and C figure 19.11 are special design spring clips. C Table 19.8 describes the ring cllips available from IFI.

19.11 Resistance Welded Fasteners

A resistance welded fastener is a part designed to be fused permanently in place by standard production welding equipment. Two popular methods of welding are used; projection and spot. With either, the fusion of the threaded fastener to a metal LM surface is the result of the natural resistance of metal to a controlled current under pressure. C Table 19.9 lists both types of fasteners available form IFI.

Resistance welding can be used with many fasteners displayed in the C figures of this textbook. For example, screws, nuts, clips and pins can be resistance welded. In projection welding, heat is localized through embossments or projections on the fastener. In spot welding, the current is directed through the entire area under the electrode tip. Welding is ususally performed by a rocker-arm type robotic spot welder. The length of the arms may range from 12 to 60 inches or more. The advantage of spot welding is that spot welders cost less than projection welders.

19.12 Chapter Summary

A lean model designer used fasteners to show the size and shape information regarding every detail of the 3-D object to be assembled. The completed model was usually an assembly display showing what type of fasteners were supplied from an approved supplier and which were unique to the lean model (LM) and must be produced or manufactured locally. Approved suppliers were members of the Industrial Fasteners Institute (IFI).

Since 1999 a LM designer can select IFI fasteners directly from the web site www.industrial-fasteners.org . In June of 1999 the Fastener Quality Act (FQA) was signed into law. This act greatly reduced the cost of world wide threads and fastener interchangeability used by American industry in its operations.

This chapter along with its illustrations was presented as an aid to proper design and display of fasteners or fastener details that can not be down loaded from the IFI. This kept the emphasis on fastener details for LM design and not fastener displays. The chapter was therefore about:

1. Permanent fasteners (welding, rivets, impact screws, etc.)
2. Assembly fasteners (keys, pins, stud bolts, etc.)
3. Application of specially designed fasteners in LM displays.

20

Lean Manufacturing Models

The essential difference between lean production modeling and lean manufacturing modeling with manufacturing software like; AUTOCAM, FABRICAM, MANUFACTURINGEXPERT, NC POLARIS, NC AUTOCODE, SMARTCAM and SURFCAM is in the use to which they are put. A LM production model is used by the engineer or technician to describe the production plan for the model geometry, construction, fabrication or the assembly of a LM. A LM manufacturing model represents what happens after the production process is decided upon. The production LM is part of an automated planning process, while the manufacturing model is part of the computer-aided machining process called CADAM . Commonly displayed manufacturing models always involve parts formation. Parts formation includes, but is not limited to the following:

1. Sand, grit and shotblasting,
2. Tumbling, Snagging and sawing,
3. Burning, cutting and welding,
4. Shaping, planing, slotting and turning,
5. Milling, drilling and profiling,
6. Hobbing, shaving and broaching,
7. Grinding, honing, lapping and polishing.

20.1 CAD Machining for Manufacture

Manufacturing CADAM must be understood by the person who prepares the lean manufacturing model (LMM). For example, a student who is enrolled in a lean modeling course at Clemson University will have a computer-aided process planning course and may also have taken a computer-aided manufacturing course as an undergraduate in mechanical engineering or engineering mechanics. In that course the following application packages would have been reviewed:

1. Manufacturing Expert from AutoDesk Inc. 800-964-6432
2. Fabricam (5 pack) from MetalSoft, Inc. 714-549-0112
3. NC polaris (11 pack) from Microproducts,   214-234-6655
4. ShopCam from Shop Systems, Inc. 219-693-9780
5. SmartCam from Point Control Co. 503-344-4470
6. SMP-81 (8 packs) from Merry Mech Inc. 612-464-8910
7. SS - (6 pack) from Striker Systems Inc. 615-672-5132
8. Surfcam (7 pack) from Surfware Inc. 818-361-5605
9. TekSoft CAD/CAM from TEKSOFT, Inc. 602-942-4982
10. Vericut from CGTECH 714-753-1050

A typical lean manufacturing model is shown in figure 20.1. In this example the model consists of a welding unit, overhead feeding station, interchangeable robot, computer control console and two work benches. The interchangeable robot head and hand gear train shown in example C (1) can be used for stud, inert gas, spot or other types of welding.

Once the head and master wrist are placed on the robot, any of the welding heads can be interchanged as shown in example C (2).

20.2 Sand, Grit and Shotblasting Units

In C figure 20.2 the welding unit has been replaced with sand, grit and shot blasting unit. This units uses a sharp dry sand directed against the work pieces on the benches by air pressure. Functions of this process include removing surface oxidation, coatings and cleanup of castings, forgings and other CAPP parts. The hoppers contain dry sand, steel grit and steel shot.

Any one or a mixture of all three can be feed through the overhead station to the robot. Functions of the mixtures include removal of parting lines, riser foot marks and peening of part surfaces to improve their fatigue life. Several changes of the robotic hands are used throughout the process as shown in C figure 20.2.

20.3 Tumbling, Snagging and Sawing Units

In C figure 20.3 the blasting unit has been replaced with a tumbling unit for removing scale, small external fins, projections, burrs and excessive tool marks from castings or rough machined parts. The parts, together with the tumbling material, is placed in a closed drum which is then rotated. The resulting agitation cleans the parts.

Snagging is handled by the robot at the benches. A hand is placed into the wrist for removing sprues, gates, risers, large fins, and projections from castings, forgings and raw materials which are too large to fit into the tumbler. Snagging is usually accomplished by rough chipping, grinding or sawing without precise limits of accuracy as shown in example C (3).

20.4 Burning, Cutting and Welding

Burning or cutting torch robot attachments are used for cutting or shaping steel sheets, rolled steel and castings by heating to extremely high temperature and simultaneously oxidizing or burning away metal with the robot hand. The hand is provided with two jets. One emits a flame combining a mixture of oxygen and accetylene for heating: the other introduces a large quantity of pure oxygen for oxidizing and burning away the metal to produce a rough cut. Welding examples were shown in C figure 20.1 and example C (4).

20.5 Shaping, Planing, Slotting and Turning

Shaping is a surface machining process for notching, key setting and facing. The cutting operation is performed by reciprocating motion (either horizontal or vertical) of a cutting tool on a machine tool called a shaper. These two types of shapers are shown in C figure 20.5 and are extremely flexible from the point of service but requires a different type of robot. The robot shown in the first four C figures is used in flexible manufacturing and is fixed to the shop floor. It has a rotating waist, interchangeable head, moveable arm through 5 axis, wrist with bend, yaw and swivel, interchangeable hands and an assortment of fingers called end effectors.

The robot shown in the remaining C figures is called a parts loader. It is not fixed to the shop floor, it is free to move about, it can have vision and object sensors. It has vertical traverse, rotational traverse, radial traverse and wrist movements as shown in example C (5).

Planing is a process similar to shaping. It is intended to produce larger, flat machined surfaces that will not fit on a shaper. A shaper has a moveable head with a cutting tool. A planer has a fixed tool head and a moveable table. Slotting can be done with either a shaper or planer and removes material to provide a relieved longitudinal area as shown in C figure 20.5. Special purpose machine tools can be used for slotting operations, they are:

1. Key seats
2. Broaches
3. Spline mills
4. milling machines

Turning is done on a computer controlled lathe as shown in example C (6). Here material is removed to produce smooth and dimensionally accurate external and internal surfaces of cylindrical, conical, shouldered or irregular form as shown in C figure 20.5. Turning operations may be per-formed castings, forgings, moldings, bars or billets. In performing turning operations, the part is rotated while the cutting tool is fed into or away from the part and is traversed along its axis of rotation. The part may be held in chucks or fixtures, or it may be supported on centers. The cutting tools are carried in cross slides or in a turret.

20.6 Milling, Drilling and Profiling

In C figure 20.6 a milling machine is used to produce internal or external surfaces of plain, complex or irregular outline to close tolerances as in example C (7). Here the milling process combines the rotation of the cutter and the feeding of the work into the path of the cutter. The cutter is supported and driven by the spindel of the machine tool. The part is supported on the machine table, which may be either numerically controlled or computer operated in 5 axes.

If either the table is stationary or the head is stationary milling is done or a drilling machine. A drilling machine, shown in example C (8) is not a drill press. It performs like a two axis milling machine. Profiling shown in C figure 20.6 is another form of two-dimensional contour milling. The travel of the cutting tool is controlled by means of a guide pin which follows the outline of a template.

20.7 Hobbing, Shaving and Broaching

Hobbing shown in C figure 20.7 is a continuous milling process. The cutter, known as a hob, and the work rotate in relation to each other on individual spindles. In addition to the rotary motion, the hob traverses across the part, or the length of the area to be hobbed. The scope of hobbing covers lean gear models manufactured from Chapter 18 and some fasteners from Chapter 19. In fact, any form or shape that is uniformly spaced on a cylindrical surface (internal or external) can be hobbed. Internal surfaces must be large enough to pass the hob through.

The upper right viewport of C figure 20.7 can also illustrate shaving or broaching as shown in example C (9). Shaving is a finishing operation which supplements hobbing to obtain a higher degree of finish, improved contour and greater accuracy of dimensions. Similar to turning, shaving can be preformed on lathes and CNC screw machines. Metal stampings from Chapter 17 are often forced through shaving dies where accuracy, surface finish and exacting contours are required. Broaching is a manufacturing process which uses a cutter known as a broach, which passes in a straight path through or over the stationary part to produce internal or external machined areas. These areas include holes of circular, square or irregular outlines, keyways and splines.

20.8 Grinding, Honing, Lapping and Polishing

Grinding may be a roughing operation like snagging, but it is generally considered a finishing operation when applied to surfaces. There are various classifications of grinding which include (a) cylindrical, (b) centerless and surface grinding as shown in example C (10).

Honing is shown in C figure 20.8 and is considered to be a finer finish than grinding. Automobile cylinders are honed during manufacture for close tolerances. Honing still has abrasive stones and therefore leave a residue on the surface to be finished. This residue can be removed by lap-ping. In lapping the stone marks are removed by a loose abrasive added to the process. It is like scrubbing a surface with a household cleaner and a sponge. Lapping may also be described as a final stock removal producing a surface quality, lean precision and accuracy. The car cylinders are ground, honed and lapped. The fuel injectors are ground, honed, lapped and polished in order to meet the mating parts requirements.

Polishing is a progressive operation performed with a set of wheels of different grain size depending upon the grit, and buffing wheels used. Grain progression and the number of grain sizes depends on the part surface and the finish desired. A mirrored polished finish is called super-finishing and requires special metals or metal coatings like chrome.

20.9 Chapter Summary

The essential difference between lean production modeling and lean manufacturing modeling with manufacturing software like; AUTOCAM, FABRICAM, MANUFACTURINGEXPERT, NC POLARIS, NC AUTOCODE, SMARTCAM and SURFCAM was in the use to which they were put. A LM production model is used by the engineer or technician to describe the production plan for the model geometry, construction, fabrication or the assembly of a LM (Lean Model). A LM manufacturing model represents what happens after the production process was decided upon. The production LM was part of an automated planning process, while the manufacturing model was part of the computer-aided machining process called CADAM . Commonly displayed manufacturing models always involve parts formation. Parts formation includes, but was not limited to the following:

1. Sand, grit and shotblasting,
2. Tumbling, Snagging and sawing,
3. Burning, cutting and welding,
4. Shaping, planing, slotting and turning,
5. Milling, drilling and profiling,
6. Hobbing, shaving and broaching,
7. Grinding, honing, lapping and polishing.

SPECIAL ASSISTANCE IS AVAILABLE ON LAPTOP CD'S FOR THE STUDY OF THE MATERIAL IN THIS REFERENCE BOOK PLEASE CONTACT THE AUTHOR @ DLR ASSOCIATES BOX 11 SUNSET, SC 29685

www.ingramcontent.com/pod-product-compliance
Lightning Source LLC
Chambersburg PA
CBHW051235050326
40689CB00007B/931